Institute of Mathematical Statistics

LECTURE NOTES – MONOGRAPH SERIES
Shanti S. Gupta, Series Editor

Volume 4

Zonal Polynomials

Akimichi Takemura

Purdue University

Institute of Mathematical Statistics
Hayward, California

Institute of Mathematical Statistics
Lecture Notes—Monograph Series
Series Editor, Shanti S. Gupta, Purdue University

The production of the IMS Lecture Notes—Monograph Series is managed by the IMS Business Office: Bruce E. Trumbo, Treasurer, and Jose L. Gonzalez, Business Manager.

Library of Congress Catalog Card Number: 84-47886

International Standard Book Number 0-940600-05-6

Copyright © 1984 Institute of Mathematical Statistics

All rights reserved

Printed in the United States of America

Contents

1 Introduction 1
2 Preliminaries on partitions and homogeneous symmetric polynomials . 7
   2.1 Partitions 7
   2.2 Homogeneous symmetric polynomials 10
3 Derivation and some basic properties of zonal polynomials 17
   3.1 Definition of zonal polynomials 17
   3.2 Integral identities involving zonal polynomials 25
   3.3 An integral representation of zonal polynomials 33
   3.4 A generating function of zonal polynomials 36
4 More properties of zonal polynomials 42
   4.1 Majorization ordering 42
   4.2 Evaluation of ${}_1\mathcal{Y}_p(I_k)$ 49
   4.3 More on integral identities 53
   4.4 Coefficients of $\mathcal{U}_q$ in $\mathcal{Y}_p$ 57
      4.4.1 Rank 1 and rank 2 cases 57
      4.4.2 Recurrence relations on the coefficients 59
   4.5 Coefficients of $\mathcal{M}_q$ 65
      4.5.1 Rank 1 and rank 2 cases 66
      4.5.2 Again on the generating function of zonal polynomials . . 67
      4.5.3 Recurrence relations of Section 4.4.2 69
      4.5.4 James' partial differential equation and recurrence relation 71
   4.6 Coefficients of $\mathcal{T}_q$ in $\mathcal{Z}_p$ 74
   4.7 Variations of the integral representation of zonal polynomials . 78
5 Complex zonal polynomials 83
   5.1 The complex normal and the complex Wishart distributions . . 84
   5.2 Derivation and properties of complex zonal polynomials 85
   5.3 Schur functions 91
   5.4 Relation between the real and the complex zonal polynomials . 95
References 98

PREFACE

This monograph presents a self-contained development of zonal polynomials in the framework of standard multivariate analysis. Except for standard tools of multivariate normal distribution theory and linear algebra, no extensive mathematics is assumed in this work. This contrasts with earlier treatments of the theory of zonal polynomials. It is hoped that the present approach makes zonal polynomials and the theory of noncentral distributions in multivariate analysis accessible to much wider audience.

This work has been based on my doctoral dissertation submitted to Stanford University in 1982. I am indebted to many people in completing my thesis and in preparing this monograph. Professor T.W. Anderson as my thesis advisor had a distinctive influence on me. Professor Ingram Olkin and Professor Persi Diaconis read the thesis and suggested many improvements. My friends Fred Huffer and Satish Iyengar read an earlier version of this work and gave very careful and useful comments. In preparing this monograph I benefited from thorough reviews by three referees. Professor Shanti Gupta, the editor of this monograph series, has been encouraging throughout the preparation of this monograph. This work was supported in part by Office of Naval Research Contract N00014-75-C-0442 and U.S. Army Research Office Contract DAAG29-82-K-0156.

Akimichi Takemura

CHAPTER 1

Introduction

In studying the behavior of power functions of various multivariate tests it is essential to develop a distribution theory for noncentral distributions. It is somewhat surprising that this is not straightforward. In fact even intuitively obvious results concerning the power functions of multivariate tests often require extensively elaborate arguments because of this difficulty. For a recent example of this see Olkin and Perlman (1980).

Let us take the noncentral χ^2 distribution and its multivariate analog, the noncentral Wishart distribution, as an example. The density of the noncentral χ^2 distribution is usually written as an infinite series. This series arises from the expansion of the exponential part of the normal density into a power series and its term by term integration with respect to irrelevant variables. In the multivariate case this integration becomes nontrivial involving an integration with respect to the Haar invariant measure on the orthogonal group. Zonal polynomials form an essential tool for studying and expressing this integration. Let us briefly review historical developments of the subject.

The first systematic studies of the noncentral Wishart distribution appeared in Anderson and Girshick (1944) and Anderson (1946). James (1955a, 1955b) introduced the integration with respect to the Haar measure on the orthogonal group explicitly and made further progress. Herz (1955) developed a theory of hypergeometric functions in matrix arguments and expressed the density of the noncentral Wishart distribution using such a function. Then James (1960, 1961a) introduced "zonal polynomials" which have a special invariance property with respect to the Haar measure and expressed the density of the noncentral Wishart distribution as an infinite series involving zonal polynomials. This infinite series provides an explicit infinite series expression for the hypergeometric function introduced by Herz. James also credits Hua (1959) for introducing zonal polynomials in another context. Zonal polynomials provided a unifying tool for the study of noncentral distributions in multivariate analysis and gave rise to an extensive literature on multivariate noncentral distributions particularly following his review article in 1964.

Examples include the distribution of the latent roots of the covariance matrix (James (1960)), the noncentral Wishart distribution (James (1961a,b)), the noncentral multivariate F distribution (James (1964)), the distribution of the roots of multivariate beta matrix (Constantine (1963)), the distribution of the canonical correlation coefficients (Constantine (1963)), the noncentral distribution of Lawley-Hotelling's trace statistic (Constantine (1966), Khatri (1967)), the noncentral distribution of Pillai's trace statistic (Khatri and Pillai (1968)), the distribution of the largest and smallest root of a Wishart or a multivariate beta distribution (Sugiyama (1966, 1967a,b), Khatri (1967, 1972), Khatri and Pillai (1968), Hayakawa (1967, 1969), Pillai and Sugiyama (1969), Krishnaiah and Chang (1971)), the distribution of quadratic functions (Khatri (1966), Hayakawa (1966, 1969), Shah (1970), Crowther and DeWaal (1973)), the distribution of multiple correlation matrix (Srivastava, 1968), the multivariate Dirichlet-type distributions (DeWaal (1970, 1972)), the distribution of some statistics for testing the equality of two covariance matrices and other hypotheses (Pillai and Nagarsenker (1972), Nagarsenker (1978)), distribution of ratio of roots (Pillai, Al-Ani, and Jouris (1969)). Pillai (1975) gives a unified treatment of several statistics.

Zonal polynomials have been found to be useful also for expressing moments of multivariate test statistics under alternative hypotheses as, for example, the moments of the generalized variance (Herz (1955)), likelihood ratio tests (Constantine (1963), Pillai, Al-Ani, and Jouris (1969)), Lawley-Hotelling's trace statistic (Constantine (1966), Khatri (1967)), the sphericity test (Pillai and

Nagarsenker (1971)), correlation matrix (DeWaal (1973)), traces of multivariate symmetric normal matrix (Hayakawa and Kikuchi (1979)). These distributional results are reviewed in Pillai (1976, 1977).

In a decision theoretic context zonal polynomials were used to evaluate risk functions in certain invariant estimation problems (Shorrock and Zidek (1976), Zidek (1978)).

In spite of the use of zonal polynomials in unifying noncentral multivariate distribution theory their theory has been considered difficult to understand. One reason for this is that James' definition of zonal polynomials (1961a) requires a rather extensive background in group representation theory. James' construction of zonal polynomials was based on fairly detailed results of classical group representation theory found in Littlewood (1950) and Weyl (1946) in particular. Compared to the extensive literature on the application of zonal polynomials, treatments of the theoretical basis of zonal polynomials have been rather scarce. Farrell (1976) and Kates (1980) in addition to James' work (1961a,b,1968) should be mentioned as advanced theoretical treatments of zonal polynomials. Farrell's construction of zonal polynomials (1976) is based on the theory of H^*-algebra found in Loomis (1953). Zonal polynomials are a particular case of spherical functions in the sense of Helgason (1962). Kates (1980) gives a thorough modern treatment of zonal polynomials in the framework of Helgason (1962) and other more recent literature on spherical functions and group representations. Unfortunately these treatments have been difficult to understand for many statisticians, including the present author. One basic reason may be that the advanced algebra required for these treatments do not form part of the usual training of mathematical statisticians.

The purpose of this monograph is to present a self-contained readable development of zonal polynomials in the framework of standard multivariate analysis. It is intended for people working with multivariate analysis. No knowledge of extensive mathematics is needed but we assume that the reader is familiar with usual multivariate analysis (e.g. Anderson (1958)) and linear algebra.

In addition to the present work there have appeared recently several elementary treatments of zonal polynomials. Let us briefly discuss these approaches and their relation to the present work. Our starting point is Saw (1977) who derived many properties of zonal polynomials using basic properties of the multivariate normal and Wishart distributions. Unfortunately at several points Saw (1977) refers to Constantine (1963) who in turn makes use of group representation theory. Therefore Saw (1977) is not entirely self-contained. Actually as Saw (1977) suggested, it turns out that only the elementary methods from Constantine (1963) are needed to complete Saw's argument. Furthermore it seems more advantageous to define zonal polynomials differently than Saw and to rearrange his logical steps; his definition of zonal polynomials appears to lack a conceptual motivation. (See Remark 3.4.1 on this point.) In our approach zonal polynomials will be defined as eigenfunctions of an expectation or integral operator. By considering the finite dimensional vector space of homogeneous symmetric polynomials of a given degree we work with vectors and matrices and define zonal polynomials simply as characteristic vectors of a certain matrix. This is done in Section 3.1.

The idea of "eigenfunctions of expected value operators" is investigated in a more general framework in recent works by Kushner and Meisner (1980) and Kushner, Lebow, and Meisner (1981). In the second paper they give a definition of zonal polynomials. They follow James' original idea but mostly use techniques of linear algebra and their approach is very helpful in understanding James' original definition. They consider the space of homogeneous polynomials of (elements of) a symmetric matrix variable A whereas we consider the space of symmetric homogeneous polynomials of the characteristic roots of A. In the former approach an extra step is needed to define zonal polynomials by requiring an orthogonal invariance. Our approach in Section 3.1 seems to be more direct.

Another welcome addition to our limited literature of elementary treatments of zonal polynomials is provided by the recent book by Muirhead (1982, Chapter 7), in which zonal polynomials are defined as eigenfunctions of a differential operator, or as solutions to partial differential equations (see James (1968, 1973)). Although Muirhead proceeds informally and many points are illustrated rather than proved, the approach can be made precise. We discuss this in Section 4.5.4. From our prejudiced viewpoint our approach seems more natural, primarily because the differential equation is hard to motivate.

What are the disadvantages of an elementary approach? One disadvantage is that often more detailed computation is needed for deriving various results than in a more abstract approach. This is in a sense a necessary tradeoff in adopting an elementary approach. However, often results in multivariate analysis require fairly heavy computation and the computation in the sequel does not seem too heavy. Another disadvantage is that the best possible results may not be obtainable. In our approach there is one important coefficient (see formula 3.4.12 below) obtained by Jamas (1961a) which we could not obtain. Furthermore our approach does not seem to apply to the recent generalization of zonal polynomials by Davis (Davis (1979, 1980, 1981), Chikuse (1981)). Except for these limitations all major properties of zonal polynomials will be proved, along with many new results. In particular the basic properties of zonal polynomials derived in Chapter 3 are sufficient for their usual applications. A good measure to judge our claim is the remarkable paper by James (1964). It is expository and contains many statements which seem to have never been proved in publication. In the sequel we will often refer to formula numbers in this paper. In summary, despite some limitations our approach seems to be very useful in removing theoretical difficulties associated with zonal polynomials.

Another difficulty of zonal polynomials is their numerical aspect. Explicit expressions for zonal polynomials are not yet known. There are several algorithms for computing their values, but they are not very fast. In Chapter 4 we study numerical aspects of zonal polynomials, especially their coefficients. Apparently some more progress is needed before zonal polynomials can be successfully applied in numerical computations.

In this monograph we will not discuss various applications of zonal polynomials in the study of multivariate noncentral distributions. Once the basic properties of zonal polynomials are established, it does not seem very difficult to express noncentral distributions in terms of zonal polynomials. Furthermore the applications are numerous as indicated above. Zonal polynomials often appear in an infinite series form which can be conveniently classified as hypergeometric function in matrix arguments (Constantine (1963), Herz (1955)). We will not discuss this topic either. Interested readers are referred to Muirhead (1982), where the noncentral distribution theory and the hypergeometric functions are amply treated. See also excellent review articles of Subrahmaniam (1976) and Pillai (1976,1977).

On the other hand we have included a development of complex zonal polynomials (zonal polynomials associated with the complex normal and the complex Wishart distributions). One reason for this is that our approach for the real case almost immediately carries over to the complex case. Another reason is that the existing theory of Schur functions provides explicit expressions for complex zonal polynomials. This becomes apparent if one compares Farrell (1980) and Macdonald (1979). See Chapter 5.

Finally let us briefly describe subsequent chapters. Chapter 2 gives preliminary material on partitions and homogeneous symmetric polynomials. Definitions and notations should be checked since they vary from book to book. In Chapter 3 we define zonal polynomials and derive their major properties. If the reader is not much interested in computational aspects of zonal polynomials the material covered in Chapter 3 should suffice for usual applications. Chapter 4 generalizes and refines the results in Chapter 3. It deals largely with computation, coefficients, etc., of zonal polynomials. The development becomes inevitably more tedious. In Chapter 5 we apply our approach to complex zonal polynomials. We show that complex zonal polynomials are the same as the Schur functions.

CHAPTER 2

Preliminaries on partitions and homogeneous symmetric polynomials

In this chapter we establish appropriate notations for partitions and homogeneous symmetric polynomials and summarize basic facts about them. They are needed for derivation of zonal polynomials in Chapter 3. It is important to check the definitions and notational conventions given in this chapter since various notational conventions on partitions and homogeneous symmetric polynomials are found in the literature. A large part of the material in this chapter is found in Macdonald (1979), Chapter 1.

§ 2.1 PARTITIONS

A set of positive integers $p = (p_1, \ldots, p_\ell)$ is called a *partition* of n if $n = p_1 + \cdots + p_\ell$. To denote p uniquely we order the elements as $p_1 \geq p_2 \geq \cdots \geq p_\ell$. $p_1, \ldots, p_\ell$ are called *parts* of p; ℓ, p_1, n are

(1)
$$\begin{aligned}\ell &= \ell(p) = length\ of\ p = number\ of\ parts, \\ p_1 &= h(p) = height\ of\ p, \\ n &= |p| = weight\ of\ p.\end{aligned}$$

respectively. The multiplicity m_i of i, $(i = 1, 2, \ldots)$ in p is defined as

(2) $$m_i = number\ of\ j\ such\ that\ p_j = i.$$

Using the m_i's p is often denoted as $p=(1^{m_1} 2^{m_2} \ldots)$. The set of all partitions of n is denoted by P_n $(=\{p : |p|=n\})$.

It is often convenient to look at p as having any number of additional zeros $p = (p_1, \ldots, p_\ell, 0, \ldots, 0)$. In this case it is understood that $p_k = 0$ for $k > \ell(p)$. With this convention addition of two partitions is defined by $(p+q)_i = p_i + q_i$, $i=1, 2, \ldots$.

A nice way of visualizing partitions is to associate the following diagrams to them. For $p = (p_1, \ldots, p_\ell)$ we associate a diagram which has p_i dots (or squares) in i-th row. For example the diagram of (4,2,2,1) is given by

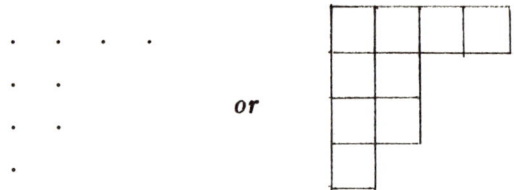

or

Figure 2.1.

We define the *conjugate partition* p' of p by means of this diagram, namely p' is a partition whose diagram is the transpose of the diagram of p. From Figure 2.1 we see $(4,2,2,1)' = (4,3,1,1)$. Clearly $p'' = (p')' = p$. Furthermore $|p| = |p'|$, $\ell(p) = h(p')$, $h(p) = \ell(p')$. More explicitly p' is determined by

(3) $$m_i(p') = p_i - p_{i+1}, \quad i = 1, \ldots, \ell.$$

Therefore for example

(4) $$\begin{aligned}\ell(p') &= m_1(p') + m_2(p') + \cdots \\ &= (p_1 - p_2) + (p_2 - p_3) + \cdots \\ &= p_1 = h(p).\end{aligned}$$

Let $s \geq h(p)$, $t \geq \ell(p)$. We define

(5) $$p^*_{s,t} = (s - p_t, s - p_{t-1}, \cdots, s - p_1).$$

From Figure 2.2 we have $(4, 2, 2, 1)^*_{4,5} = (4, 3, 2, 2, 0)$. Note that

(6) $$|p^*_{s,t}| = st - |p|.$$

§ 2.1 Partitions

```
                    s
                 .  .  .  .
                 .  .  x  x
        t        .  .  x  x
                 .  x  x  x
                 x  x  x  x
```

Figure 2.2.

Now we introduce two orderings in P_n. The first one is called the *lexicographic ordering* ($>$). In this ordering p is said to be *higher than* q ($p > q$) if

(7) $$p_1 = q_1, \ldots, p_{k-1} = q_{k-1}, p_k > q_k \quad \text{for some } k.$$

This is a total ordering. For example P_4 is ordered as $(4) > (3,1) > (2,2) > (2,1,1) > (1,1,1,1)$.

This ordering is preserved by addition.

Lemma 1. *If $p^1 \geq q^1$, $p^2 \geq q^2$ then $p^1 + p^2 \geq q^1 + q^2$ with equality iff $p^1 = q^1, p^2 = q^2$.*

Proof is easy and omitted.

Another ordering is the *majorization* ordering. p *majorizes* q ($p \succ q$) if and only if

(8) $$p_1 \geq q_1, \; p_1 + p_2 \geq q_1 + q_2, \; \ldots, \; p_1 + \cdots + p_k \geq q_1 + \cdots + q_k, \ldots$$

Note that for $k \geq max(\ell(p), \ell(q))$ the equality holds because both sides are equal to the weight n. Majorization is a partial ordering and it is stronger than the lexicographic ordering:

Lemma 2. *If $p \succ q$ then $p \geq q$.*

Proof. Suppose $p_1 = q_1, \cdots, p_{k-1} = q_{k-1}, p_k \neq q_k$. Then $p_1 + \cdots + p_k \geq q_1 + \cdots + q_k$ implies $p_k > q_k$. Hence $p > q$. ∎

Remark 1. The converse of Lemma 2 is false. For example $(3,1,1,1) > (2,2,2)$ but there is no majorization between these two.

Analogous to Lemma 1 we have

Lemma 3. *If $p^1 \succ q^1$, $p^2 \succ q^2$, then $p^1 + p^2 \succ q^1 + q^2$ with equality iff $p^1 = q^1$, $p^2 = q^2$.*

Proof. For any k

$$(p_1^1 + p_1^2) + \cdots + (p_k^1 + p_k^2) \geq (q_1^1 + q_1^2) + \cdots + (q_k^1 + q_k^2)$$

with equality iff $p_1^i + \cdots + p_k^i = q_1^i + \cdots + q_k^i$, $i = 1, 2$. ∎

The last lemma in this section is the following:

Lemma 4. *Let $p, q \in \mathcal{P}_n$ and let s, t be such that $s \geq h(p), s \geq h(q), t \geq \ell(p), t \geq \ell(q)$. Then $p \succ q$ if and only if $p_{s,t}^* \succ q_{s,t}^*$.*

Proof. (8) holds if and only if $p_1 - n \geq q_1 - n$, $p_1 + p_2 - n \geq q_1 + q_2 - n, \ldots$. Noting that $n = p_1 + \cdots + p_t = q_1 + \cdots + q_t$, these inequalities in the reversed order imply $p_{s,t}^* \succ q_{s,t}^*$.

§ 2.2 HOMOGENEOUS SYMMETRIC POLYNOMIALS

Let $f(x_1, \ldots, x_k)$ be a polynomial in $x_1, \ldots, x_k$. f is *homogeneous (of degree n)* if f has only n-th degree terms. f is *symmetric* if

$$(1) \qquad f(x_1, \ldots, x_k) = f(x_{i_1}, \ldots, x_{i_k}),$$

where $(i_1, \ldots, i_k)$ is any permutation of $(1, \ldots, k)$. Let V_n denote the set of all n-th degree homogeneous symmetric polynomials including the constant $f \equiv 0$. We look at V_n as a vector space where addition is the usual addition of polynomials. Let $f \in V_n$ and suppose that f has a term $ax_1^{p_1} \cdots x_\ell^{p_\ell}$ $((p_1, \ldots, p_\ell) \in \mathcal{P}_n)$, then by symmetry it also has a term $ax_{i_1}^{p_1} \cdots x_{i_\ell}^{p_\ell}$ where $i_1, \ldots, i_\ell$ are distinct integers taken from $(1, \ldots, k)$. Counting all different terms we see that

f can be written as a linear combination of *monomial symmetric functions* M_p, $p \in P_n$,

(2) $$f = \sum_{p \in P_n} a_p M_p,$$

where

(3) $$M_p(x_1, \ldots, x_k) = \sum_{(i_1, \ldots, i_\ell) \subset (1, \ldots, k)} x_{i_1}^{p_1} \cdots x_{i_\ell}^{p_\ell}.$$

In (3) we count only distinguishable terms. For example

(4) $$M_{(1,1)} = \sum_{i<j} x_i x_j.$$

Sometimes it is more convenient to use *augmented monomial symmetric function* $\mathcal{A}M_p$ for which the summation in (3) is over all permutations of ℓ different integers from $(1, \ldots, k)$. Therefore for example

(5) $$\mathcal{A}M_{(1,1)} = \sum_{i \neq j} x_i x_j = 2 M_{(1,1)}.$$

In general

(6) $$\mathcal{A}M_p = \left(\prod_{i=1}^{h(p)} m_i! \right) M_p.$$

where $(p_1, \ldots, p_\ell) = (1^{m_1} 2^{m_2} \ldots)$.

We note that in (2) the number of variables k does not play an explicit role. Actually M_p can be defined for any number of variables by (3) and

(7) $$M_p(x_1, \ldots, x_k, 0, \ldots, 0) = M_p(x_1, \ldots, x_k).$$

Hence it suffices to consider M_p which is defined for sufficiently large number of variables. Now suppose

(8) $$\sum_{p \in P_n} a_p M_p = 0.$$

We look at terms of the form $x_1^{q_1} \cdots x_\ell^{q_\ell}$. Differentiating (8) p_i times with respect to x_i, $i=1,\ldots,\ell$ we have $(\prod p_i!)a_p = 0$. Hence $a_p = 0$ for all $p \in P_n$ and $M_p, p \in P_n$ are linearly independent in V_n. (Of course if $k < \ell(p)$ then $M_p(x_1,\ldots,x_k) = 0$ which is linearly dependent in a trivial sense. But as above we consider k to be sufficiently large. For more detail see Section 4.1.) From (2) and (8) it follows that $\{M_p,\ p \in P_n\}$ forms a basis of V_n. This is a rather obvious basis. We want to consider other bases. The following lemma is useful for this purpose.

Lemma 1. *If A is an upper triangular matrix with nonzero diagonal elements, then A^{-1} has the same property. Furthermore if A has diagonal elements 1 and integral offdiagonal elements, then A^{-1} has the same property.*

Proof. The first statement is obvious. For the second statement note $|A|=1$. Hence $A^{-1} = (a^{ij}) = (\Delta_{ji})$, where Δ_{ij} is a cofactor of A. But Δ_{ij}'s are integers. ∎

Now we consider products of elementary symmetric functions. Let

$$(9) \qquad u_r = \sum_{i_1 < \cdots < i_r} x_{i_1} \cdots x_{i_r}$$

be the r-th elementary symmetric function. For $p \in P_n$ we define

$$(10) \qquad \mathcal{U}_p = u_1^{p_1-p_2} u_2^{p_2-p_3} \cdots u_\ell^{p_\ell}.$$

The degree of $\mathcal{U}_p$ is

$$(11) \qquad (p_1 - p_2) + 2(p_2 - p_3) + \cdots + \ell p_\ell = p_1 + \cdots + p_\ell = n.$$

Hence $\mathcal{U}_p \in V_n$. $\mathcal{U}_p$ defined by (10) corresponds to $\mathcal{U}_{p'}$ in Macdonald's notation (1979).

Lemma 2.

$$(12) \qquad \mathcal{U}_p = M_p + \sum_{q < p} a_{pq} M_q,$$

§ 2.2 Homogeneous symmetric polynomials

where a_{pq} are integers.

Proof. Consider monomial terms of the form $x_1^{q_1} x_2^{q_2} \cdots x_k^{q_k}$, $q = (q_1, \ldots, q_k) \in \mathcal{P}_n$. Now

$$\mathcal{U}_p = (x_1 + \cdots)^{p_1 - p_2}(x_1 x_2 + \cdots)^{p_2 - p_3} \cdots (x_1 \cdots x_\ell + \cdots)^{p_\ell}.$$

Hence the highest order term obtained by expanding $\mathcal{U}_p$ is

$$x_1^{p_1-p_2}(x_1 x_2)^{p_2-p_3} \cdots (x_1 \cdots x_\ell)^{p_\ell} = x_1^{p_1} x_2^{p_2} \cdots x_\ell^{p_\ell},$$

which has coefficient 1. It is clear that other terms are lower in the lexicographic ordering and have integral coefficients. ∎

Remark 1. For a stronger result see Lemma 4.1.1.

We order $\mathcal{M}_p, \mathcal{U}_p$, $p \in \mathcal{P}_n$ according to the lexicographic ordering and form two vectors:

$$(13) \qquad \mathcal{M} = \begin{pmatrix} \mathcal{M}_{(n)} \\ \mathcal{M}_{(n-1,1)} \\ \vdots \\ \mathcal{M}_{(1^n)} \end{pmatrix}, \qquad \mathcal{U} = \begin{pmatrix} \mathcal{U}_{(n)} \\ \mathcal{U}_{(n-1,1)} \\ \vdots \\ \mathcal{U}_{(1^n)} \end{pmatrix}.$$

Then Lemma 2 implies that

$$(14) \qquad \mathcal{U} = \mathbf{A}\mathcal{M}, \qquad \mathbf{A} = (a_{pq}),$$

where $\mathbf{A}$ is a matrix satisfying the condition of Lemma 1. Therefore considering $\mathbf{A}^{-1} = (a^{pq})$ we obtain

$$(15) \qquad \mathcal{M}_p = \mathcal{U}_p + \sum_{q < p} a^{pq} \mathcal{U}_q,$$

where a^{pq} are integers. We see that $\{\mathcal{U}_p, p \in \mathcal{P}_n\}$ forms another basis of V_n.

Product of $\mathcal{U}$ functions corresponds to the addition of partitions.

Lemma 3.

$$\mathcal{U}_p \mathcal{U}_q = \mathcal{U}_{p+q}. \tag{16}$$

Proof is easy and omitted.

The third basis of V_n is given by product of power sums. Let

$$t_r = \sum x_i^r. \tag{17}$$

For $p \in \mathcal{P}_n$ we define

$$\mathcal{T}_p = t_1^{p_1-p_2} t_2^{p_2-p_3} \cdots t_\ell^{p_\ell}. \tag{18}$$

$\mathcal{T}_p$ defined by (18) corresponds to $\mathcal{T}_{p'}$ in Macdonald (1979) and in Saw (1977). Here we prefer the above definition because of the simpler relation between $\mathcal{U}_p$ and $\mathcal{T}_p$.

Let

$$U(s) = \prod(1 + sx_i) = 1 + u_1 s + u_2 s^2 + \ldots \tag{19}$$

be a generating function of u's. Then

$$\begin{aligned}\log U(s) &= \sum \log(1 + sx_i) \\ &= st_1 - \frac{s^2}{2} t_2 + \cdots + (-1)^{r-1} \frac{s^r}{r} t_r + \cdots. \end{aligned} \tag{20}$$

On the other hand

$$\log U(s) = (u_1 s + u_2 s^2 + \cdots) - \frac{1}{2}(u_1 s + u_2 s^2 + \cdots)^2 + \cdots. \tag{21}$$

Comparing coefficients of s^r in (20) and (21) we see

$$\begin{aligned} t_r &= (-1)^{r-1} r \left\{ u_r + \sum_{q>(1^r), q \in \mathcal{P}_r} a_{rq} \mathcal{U}_q \right\} \\ &= (-1)^{r-1} r \left\{ \mathcal{U}_{(1^r)} + \sum_{q>(1^r), q \in \mathcal{P}_r} a_{rq} \mathcal{U}_q \right\}. \end{aligned} \tag{22}$$

Actually

(23)
$$a_{rq} = \frac{(-1)^{q_1-1}}{q_1}\binom{q_1}{q_1-q_2, q_2-q_3, \ldots, q_{\ell(q)}}$$
$$= \frac{(-1)^{q_1-1}(q_1-1)!}{(q_1-q_2)!\cdots q_{\ell(q)}!}.$$

This follows from the fact that $\mathcal{U}_q$ being a product of q_1 elementary symmetric functions comes only from the q_1-th power term in the expansion of log in (21).

Now

(24)
$$\mathcal{T}_p = \prod_{r=1}^{\ell(p)} t_r^{p_r-p_{r+1}}$$
$$= \prod_{r=1}^{\ell(p)}\left[(-1)^{r-1}r\{\mathcal{U}_{(1^r)} + \sum_{q>(1^r), q\in P_r} a_{rq}\mathcal{U}_q\}\right]^{p_r-p_{r+1}}.$$

By Lemma 2.1.1 and Lemma 3 the lowest order term in (24) is given by

(25)
$$\prod_{r=1}^{\ell}[(-1)^{r-1}r\mathcal{U}_{(1^r)}]^{p_r-p_{r+1}}$$
$$= \prod_{r=1}^{\ell}[(-1)^{r-1}r]^{p_r-p_{r+1}} u_1^{p_1-p_2} u_2^{p_2-p_3}\cdots u_\ell^{p_\ell}$$
$$= (-1)^{|p|-p_1}\left(\prod_{r=1}^{\ell} r^{p_r-p_{r+1}}\right)\mathcal{U}_p.$$

Hence

Lemma 4.

(26)
$$\mathcal{T}_p = \sum_{q\geq p} a_{pq}\mathcal{U}_q,$$

where

(27)
$$a_{pp} = (-1)^{|p|-p_1}\prod_{r=1}^{\ell(p)} r^{p_r-p_{r+1}} \neq 0.$$

Let
$$\mathcal{T} = \begin{pmatrix} \mathcal{T}_{(n)} \\ \mathcal{T}_{(n-1,1)} \\ \cdot \\ \cdot \\ \cdot \\ \mathcal{T}_{(1^n)} \end{pmatrix}.$$

Then Lemma 4 shows that

(28) $$\mathcal{T} = F\mathcal{U},$$

where F is lower triangular with nonzero diagonal elements. Hence $\{\mathcal{T}_p, p \in \mathcal{P}_n\}$ forms a basis of V_n.

Remark 2. To show that $\{\mathcal{T}_p, p \in \mathcal{P}_n\}$ is a basis it is much easier to note

$$\mathcal{T}_{p'} = \mathcal{A}M_p + \sum_{q>p} a_{pq} \mathcal{A}M_q,$$

where a_{pq} are integers. But we will use Lemma 4 in Section 4.6.

We study symmetric functions further in Section 4.1 and Section 5.3. However the material covered so far suffices to derive zonal polynomials which form another basis of V_n.

Remark 3. For the coefficients of basis functions we generally use $a_p, b_p, \ldots, a_{pq}, b_{pq}$, etc. Since there are many instances of this, it is impossible to use different symbols for each case. For example a_{pq} in Lemma 2 and in Lemma 4 are different.

CHAPTER 3

Derivation and some basic properties of zonal polynomials

In this chapter we define (real) zonal polynomials and derive their basic properties. The results derived in this chapter are sufficient for usual applications of zonal polynomials. Some remarks on notation seem appropriate here. We define zonal polynomials as characteristic vectors of a certain linear transformation r from V_n to V_n. The normalization is rather arbitrary for a characteristic vector and many properties of zonal polynomials are independent of particular normalization. Corresponding to different normalizations, different symbols such as Z_p, C_p have been used to denote zonal polynomials. We find it advantageous to use still another normalization in addition to those corresponding to Z_p, C_p. Considering these circumstances we use $\mathcal{Y}_p$ for an unnormalized zonal polynomial. $_1\mathcal{Y}_p$ is used to denote a zonal polynomial normalized so that the coefficient of $\mathcal{U}_p$ or $\mathcal{M}_p$ is 1.

§ 3.1 DEFINITION OF ZONAL POLYNOMIALS

As mentioned earlier we define zonal polynomials as characteristic vectors of a certain matrix. The matrix in question will be triangular and we begin by a lemma concerning a triangular matrix and its characteristic vectors.

Lemma 1. *Let $T = (t_{ij})$ be an $n \times n$ upper triangular matrix with distinct* diagonal elements. Let $\Lambda = \text{diag}(t_{11}, \ldots, t_{nn})$. Then there exists a nonsingular upper triangular matrix B satisfying

(1) $$BT = \Lambda B.$$

B is uniquely determined up to a (possibly different) multiplicative constant for each row.

Proof is straightforward and omitted. Note that t_{ii}, $i = 1, \ldots, n$ are characteristic roots of T and i-th row of B is the characteristic vector (from the left) associated with t_{ii}.

Remark 1. This lemma seems to be well known to people in numerical analysis although an explicit reference is not easy to find. It is very briefly mentioned on page 365 of Stewart (1973) in connection with the QR algorithm. The QR algorithm is designed to transform a general matrix to a triangular form in order to obtain the characteristic roots and vectors.

For a $k \times k$ matrix $A = (a_{ij})$ we denote its (possibly complex) characteristic roots by

(2) $$\alpha = (\alpha_1, \ldots, \alpha_k) = \lambda(A),$$

and (the determinant of) a principal minor by

(3) $$A(i_1, \ldots, i_\ell) = \begin{vmatrix} a_{i_1 i_1} & \cdots & a_{i_1 i_\ell} \\ \vdots & & \vdots \\ a_{i_\ell i_1} & \cdots & a_{i_\ell i_\ell} \end{vmatrix}.$$

For a matrix argument we define

$$\mathcal{U}_p(A) = \mathcal{U}_p(\alpha) = \mathcal{U}_p(\lambda(A)).$$

As is easily seen by expanding the determinant $|A - \lambda I|$ the r-th elementary symmetric function of the roots of a matrix A is equal to the sum of $r \times r$ principal minors, namely

(4) $$\mathcal{U}_{(1^r)}(A) = u_r(\alpha_1, \ldots, \alpha_k) = \sum_{i_1 < \cdots < i_r} A(i_1, \ldots, i_r).$$

§ 3.1   *Definition of zonal polynomials*

(See Theorem 7.1.2 of Mirsky (1955) for example.) Hence

$$(5) \qquad \mathcal{U}_p(A) = \{\sum_{i_1} A(i_1)\}^{p_1-p_2} \{\sum_{i_1<i_2} A(i_1,i_2)\}^{p_2-p_3} \cdots.$$

Note that (4) and (5) holds for general (not necessarily symmetric) matrix A.

Now let A be symmetric and consider a (linear) transformation $\tau_\nu : V_n \to V_n$ defined by

$$(6) \qquad (\tau_\nu(\mathcal{U}_p))(A) = (\tau_\nu \mathcal{U}_p)(A) = \mathcal{E}_W\{\mathcal{U}_p(AW)\},$$

where W is a random symmetric matrix having a Wishart distribution $\mathcal{W}(I_k, \nu)$, $\nu \geq k$. Here $\mathcal{W}(\Sigma, \nu)$ denotes the Wishart distribution with covariance Σ and degrees of freedom ν. (τ_ν is defined for the basis $\{\mathcal{U}_p\}$ by (6) and for general elements of $V_n(A)$ τ_ν is given by the linearity of expectation.) First we need to verify:

Lemma 2. $\tau_\nu \mathcal{U}_p \in V_n$.

Proof.   Since A is symmetric it can be written as $A = \Gamma D \Gamma'$ where Γ is orthogonal and $D = \text{diag}(\alpha_1, \ldots, \alpha_k)$. Now $\mathcal{U}_p(AW) = \mathcal{U}_p(\Gamma D \Gamma' W) = \mathcal{U}_p(D\Gamma' W\Gamma)$ because the nonzero roots are invariant when the matrices are permuted cyclically. Since the distribution of $\Gamma' W \Gamma$ is the same as the distribution of W, we can take $A = \text{diag}(\alpha_1, \ldots, \alpha_k)$ without loss of generality. Then

$$(7) \qquad AW(i_1, \ldots, i_r) = (\alpha_{i_1} \cdots \alpha_{i_r})W(i_1, \ldots, i_r).$$

For example

$$(8) \qquad AW(1,2) = \begin{vmatrix} \alpha_1 w_{11} & \alpha_1 w_{12} \\ \alpha_2 w_{21} & \alpha_2 w_{22} \end{vmatrix} = \alpha_1 \alpha_2 \begin{vmatrix} w_{11} & w_{12} \\ w_{21} & w_{22} \end{vmatrix}.$$

From (4) and (7) the *r-th* elementary symmetric function of the characteristic roots of AW can be written as

$$(9) \qquad u_r(\lambda(AW)) = \sum_{i_1<\cdots<i_r} \alpha_{i_1} \cdots \alpha_{i_r} W(i_1, \ldots, i_r).$$

Substituting this into (5) and taking the expectation we obtain

$$(10) \qquad (\tau_\nu \mathcal{U}_p)(A) = \mathcal{E}_W(\sum_{i_1} \alpha_{i_1} W(i_1))^{p_1-p_2} (\sum_{i_1<i_2} \alpha_{i_1}\alpha_{i_2} W(i_1,i_2))^{p_2-p_3} \cdots.$$

Clearly this belongs to V_n.   ∎

τ_ν has the following triangular property.

Corollary 1.

$$(11) \qquad (\tau_\nu \mathcal{U}_p)(A) = \lambda_{\nu p} \mathcal{U}_p(A) + \sum_{q < p} a_{pq} \mathcal{U}_q(A).$$

Proof. It suffices to show this for a diagonal matrix $A = \mathrm{diag}(\alpha_1, \ldots, \alpha_k)$. As in the proof of Lemma 2.2.2 the highest monomial term in (10) is of the form

$$(12) \qquad \alpha_1^{p_1} \alpha_2^{p_2} \cdots \alpha_\ell^{p_\ell} \mathcal{E}_W \left\{ W(1)^{p_1 - p_2} W(1,2)^{p_2 - p_3} \cdots W(1, \ldots, \ell)^{p_\ell} \right\}.$$

Then using (2.2.15) we see that $(\tau_\nu \mathcal{U}_p)(A)$ expressed as a linear combination of $\mathcal{U}_q$'s involves only q's such that $q \leq p$. In particular the leading coefficient is

$$(13) \qquad \lambda_{\nu p} = \mathcal{E}_W \left\{ W(1)^{p_1 - p_2} W(1,2)^{p_2 - p_3} \cdots W(1, \ldots, \ell)^{p_\ell} \right\}.$$

∎

Remark 2. The constants a_{pq} in (11) depend on the degrees of freedom ν.

Remark 3. To be complete we have to verify that (6) does not depend on the number of variables k or more precisely we need to verify

$$(14) \qquad (\tau_\nu \mathcal{U}_p)(\alpha_1, \ldots, \alpha_k, 0, \ldots, 0) = (\tau_\nu \mathcal{U}_p)(\alpha_1, \ldots, \alpha_k),$$

for any number (m) of additional zeros. Note that the left hand side is defined using expectation with respect to $\mathcal{W}(I_{k+m}, \nu)$. Now recall that the marginal distribution of the $k \times k$ upper left hand corner of $\mathcal{W}(I_{k+m}, \nu)$ is $\mathcal{W}(I_k, \nu)$ and (10) depends only on the $k \times k$ upper left hand corner of the Wishart matrix. Therefore (14) holds.

By Corollary 1 τ_ν expressed in an appropriate matrix form is an upper triangular matrix. In order to apply Lemma 1 we want to evaluate the "diagonal elements" $\lambda_{\nu p}$ in (13). For that purpose we use the following well known result.

Lemma 3. Let W be distributed according to $\mathcal{W}(I_k, \nu)$. Let $T = (t_{ij})$ be a lower triangular matrix with nonnegative diagonal elements such that $W = TT'$. Then t_{ij}, $i \geq j$, are independently distributed as $t_{ij} \sim \mathcal{N}(0,1)$, $i > j$, $t_{ii} \sim \chi(\nu - i + 1)$ where $\chi(\nu - i + 1)$ denotes the chi-distribution with $\nu - i + 1$ degrees of freedom.

For a proof see Wijsman (1959) or Kshirsagar (1959).

Corollary 2.

$$
\begin{aligned}
\lambda_{\nu p} &= 2^n \prod_{i=1}^{\ell} \Gamma[p_i + \tfrac{1}{2}(\nu + 1 - i)] / \Gamma[\tfrac{1}{2}(\nu + 1 - i)] \\
&= 2^n \prod_{i=1}^{\ell} \left(\frac{\nu + 1 - i}{2} \right)_{p_i} \\
&= \nu(\nu + 2) \cdots (\nu + 2(p_1 - 1)) \\
&\quad \cdot (\nu - 1)(\nu + 1) \cdots (\nu - 1 + 2(p_2 - 1)) \\
&\quad \cdots \\
&\quad \cdot (\nu - \ell + 1) \cdots (\nu - \ell + 1 + 2(p_\ell - 1)),
\end{aligned}
$$
(15)

where $\ell = \ell(p)$ and $(a)_k = a(a+1) \cdots (a + k - 1)$.

Proof. Note

(16) $$W(1, \ldots, r) = (t_{11} \cdots t_{rr})^2.$$

Substituting this into (13) we obtain

(17) $$\lambda_{\nu p} = \mathcal{E}\{t_{11}^{2p_1} t_{22}^{2p_2} \cdots t_{\ell\ell}^{2p_\ell}\}.$$

Now t_{ii}^2 is distributed according to $\chi^2(\nu - i + 1)$ and $\mathcal{E} t_{ii}^{2p_i} = (\nu - i + 1)(\nu - i + 3) \cdots (\nu - i + 1 + 2(p_i - 1))$. From this we obtain (15). ∎

This proof is given in Constantine (1963) in a slightly different form.

Using the vector notation introduced in (2.2.13) let

(18) $$\tau_\nu(\mathcal{U}) = \begin{pmatrix} \tau_\nu(\mathcal{U}_{(n)}) \\ \tau_\nu(\mathcal{U}_{(n-1,1)}) \\ \cdot \\ \cdot \\ \cdot \\ \tau_\nu(\mathcal{U}_{(1^n)}) \end{pmatrix}.$$

Then Corollary 1 shows that

(19) $$\tau_\nu(\mathcal{U}) = \boldsymbol{T}_\nu \mathcal{U},$$

where $\boldsymbol{T}_\nu$ is an upper triangular matrix with diagonal elements $t_{pp} = \lambda_{\nu p}$. $\boldsymbol{T}_\nu$ almost fits the condition of Lemma 1. The question now is what ν to take. Actually a particular choice of ν does not matter; we have:

Lemma 4. *There exists a nonsingular upper triangular matrix $\boldsymbol{B}$ such that*

(20) $$\boldsymbol{B}\boldsymbol{T}_\nu = \boldsymbol{\Lambda}_\nu \boldsymbol{B} \quad \text{for all } \nu,$$

where $\boldsymbol{\Lambda}_\nu = \mathrm{diag}(\lambda_{\nu p}, p \in \mathcal{P}_n)$. $\boldsymbol{B}$ is uniquely determined up to a (possibly different) multiplicative constant for each row.

Lemma 4 shows that $\boldsymbol{T}_\nu$ has the same set of characteristic vectors (from the left) for all ν. A proof of this will be given later in this section. Now we define zonal polynomials using this $\boldsymbol{B}$.

Definition 1. (*zonal polynomials*)

Let $\boldsymbol{B}$ be as in Lemma 4. Zonal polynomials $\mathcal{Y}_p$, $p \in \mathcal{P}_n$, are defined by

(21) $$\mathcal{Y} = \begin{pmatrix} \mathcal{Y}_{(n)} \\ \mathcal{Y}_{(n-1,1)} \\ \cdot \\ \cdot \\ \cdot \\ \mathcal{Y}_{(1^n)} \end{pmatrix} = \boldsymbol{B}\mathcal{U}.$$

Remark 4. $\boldsymbol{B}$ is upper triangular and therefore $\mathcal{Y}_p$ is a linear combination of $\mathcal{U}_q$'s (or $\mathcal{M}_q$'s) with $q \le p$. It follows that $\{\mathcal{Y}_p, p \in \mathcal{P}_n\}$ forms a basis of V_n.

Remark 5. Since each row of $\boldsymbol{B}$ is determined uniquely up to a multiplicative constant $\mathcal{Y}_p$ is determined up to normalization. We use $\mathcal{Y}_p$ to denote an unnormalized zonal polynomial.

In order to prove Lemma 4 we first establish that the $\boldsymbol{T}_\nu$'s commute with each other.

§ 3.1 *Definition of zonal polynomials* 23

Lemma 5.

(22) $$T_\nu T_\mu = T_\mu T_\nu.$$

Proof. For a symmetric positive semi-definite matrix A let $A^{\frac{1}{2}}$ be the symmetric positive semi-definite square root, i.e., $A^{\frac{1}{2}} = \Gamma D^{\frac{1}{2}} \Gamma'$ where Γ is orthogonal and D is diagonal in $A = \Gamma D \Gamma'$. Now let W, V be independently distributed according to $\mathcal{W}(I_k, \nu)$, $\mathcal{W}(I_k, \mu)$ respectively. Consider

(23) $$\mathcal{E}_{W,V}\{\mathcal{U}(A^{\frac{1}{2}} V A^{\frac{1}{2}} W)\},$$

where $\mathcal{U} = (\mathcal{U}_{(n)}, \mathcal{U}_{(n-1,1)}, \ldots, \mathcal{U}_{(1^n)})'$. Taking expectation with respect to W first we obtain

(24) $$\begin{aligned}\mathcal{E}_{W,V}&\{\mathcal{U}(A^{\frac{1}{2}} V A^{\frac{1}{2}} W)\} \\ &= \mathcal{E}_V\{T_\nu \mathcal{U}(A^{\frac{1}{2}} V A^{\frac{1}{2}})\} \\ &= \mathcal{E}_V\{T_\nu \mathcal{U}(AV)\} \\ &= T_\nu \mathcal{E}_V\{\mathcal{U}(AV)\} \\ &= T_\nu T_\mu \mathcal{U}(A).\end{aligned}$$

We used the cyclic permutation of the matrices since nonzero characteristic roots are invariant. Similarly taking expectation with respect to V first we obtain

(25) $$\mathcal{E}_{W,V}\{\mathcal{U}(A^{\frac{1}{2}} V A^{\frac{1}{2}} W)\} = T_\mu T_\nu \mathcal{U}(A).$$

Hence $T_\nu T_\mu \mathcal{U}(A) = T_\mu T_\nu \mathcal{U}(A)$ for any symmetric positive semidefinite A. Now a polynomial is identically equal to zero if it is zero for all nonnegative arguments. This implies $T_\nu T_\mu = T_\mu T_\nu$. ∎

See Theorem 2.2 of Kushner, Lebow, and Meisner (1981) for an analogous result in a more general framework.

Now we give a proof of Lemma 4.

Proof of Lemma 4: Consider $\lambda_{\nu p}$ given by (15). Let us look at $\lambda_{\nu p}$ as a polynomial in ν. They are different polynomials for different partitions since they have different sets of roots. Now two different polynomials can match only finite number of times. It follows that for a sufficiently large ν_0, $\lambda_{\nu_0 p}$, $p \in P_n$, are all different. Let ν_0 be fixed such that $\lambda_{\nu_0 p}, p \in P_n$ are all different. Let B be the matrix in (1) with T replaced by T_{ν_0}. Note that the uniqueness part of Lemma 4 is already established now. Let $\Lambda = \text{diag}(\lambda_{\nu_0 p}, p \in P_n)$. Then for any μ $\Lambda(BT_\mu) = (\Lambda B)T_\mu = (BT_{\nu_0})T_\mu = B(T_{\nu_0}T_\mu) = B(T_\mu T_{\nu_0}) = (BT_\mu)T_{\nu_0}$, or $\Lambda B_1 = B_1 T_{\nu_0}$ where $B_1 = BT_\mu$. Now by the uniqueness part of Lemma 1 we have $B_1 = DB$ for some diagonal D or $BT_\mu = DB$. Considering the diagonal elements we see that $D = \Lambda_\mu = \text{diag}(\lambda_{\mu p}, p \in P_n)$. Therefore $BT_\mu = \Lambda_\mu B$ for all μ. ∎

We defined zonal polynomials by defining their coefficients. From a little bit more abstract viewpoint they are eigenfunctions of the linear operator τ_ν and the results in this section can be summarized as follows.

Theorem 1. *Let $\mathcal{Y}_p$ be a zonal polynomial then*

(26) $$(\tau_\nu \mathcal{Y}_p)(A) = \mathcal{E}_W \mathcal{Y}_p(AW) = \lambda_{\nu p} \mathcal{Y}_p(A),$$

where $W \sim \mathcal{W}(I_k, \nu)$, A is symmetric and $\lambda_{\nu p}$ is given in (15). Conversely (26) (for all sufficiently large ν and for all symmetric A) implies that $\mathcal{Y}_p$ is a zonal polynomial.

Proof. $\mathcal{Y} = (\mathcal{Y}_{(n)}, \mathcal{Y}_{(n-1,1)}, \ldots, \mathcal{Y}_{(1^n)})' = B\mathcal{U}$. Hence by Lemma 4

(27) $$\begin{aligned} \mathcal{E}_W\{\mathcal{Y}(AW)\} &= \mathcal{E}_W\{B\mathcal{U}(AW)\} \\ &= B\mathcal{E}_W\{\mathcal{U}(AW)\} \\ &= BT_\nu \mathcal{U}(A) \\ &= \Lambda_\nu B\mathcal{U}(A) \\ &= \Lambda_\nu \mathcal{Y}(A). \end{aligned}$$

Therefore (26) holds. Conversely assume (26). Let $\mathcal{Y}_p = \sum_{q \in P_n} a_q \mathcal{U}_q$. Then (26) implies

$$a' T_\nu = \lambda_{\nu p} a',$$

where $a' = (a_{(n)}, \ldots, a_{(1^n)})$. Now by the uniqueness part of Lemma 4 a' coincides with the "p-th" row of B up to a multiplicative constant. Therefore $\mathcal{Y}_p$ is a zonal polynomial. ∎

Corollary 3.

(28) $$\mathcal{E}_W \mathcal{Y}_p(AW) = \lambda_{\nu p} \mathcal{Y}_p(A\Sigma),$$

where A is symmetric and W is distributed according to $\mathcal{W}(\Sigma, \nu)$.

Proof. This follows from (26) noting that if $W = \Sigma^{\frac{1}{2}} W_1 \Sigma^{\frac{1}{2}}$ then W_1 is distributed according to $\mathcal{W}(I_k, \nu)$. ∎

The converse part of Theorem 1 will be strengthened in Theorem 4.1.2 and will be used to show that a particular symmetric polynomial is a zonal polynomial. See Sec. 3.3, Sec. 4.4, and Sec. 4.7.

Remark 6. In (21) zonal polynomials are defined as linear combinations of $\mathcal{U}_q$'s. Therefore by (4) and (5) $\mathcal{Y}_p(A)$ makes sense even when A is not symmetric. This has been already used in the form $\mathcal{Y}_p(AW)$ in (26). This might be slightly confusing because τ_ν (hence T_ν and $\mathcal{B}$) was defined by considering only symmetric matrices. Indeed in most cases arguments for zonal polynomials are symmetric matrices.

§ 3.2 INTEGRAL IDENTITIES INVOLVING ZONAL POLYNOMIALS

In addition to (3.1.26) the zonal polynomials satisfy other integral identities. The fundamental one (Theorem 1 below) is related to the uniform distribution of orthogonal matrices. The idea of "averaging with respect to the uniform distribution of orthogonal matrices" or "averaging over orthogonal group" was a very important idea of James for the motivation of introducing the zonal polynomials.

A random orthogonal matrix H is said to have the *Haar invariant* distribution or the *uniform distribution* if the distribution of $H\Gamma$ is the same for every orthogonal Γ. More formally, a probability measure P on the Borel field of orthogonal matrices is *Haar invariant* if

(1) $$P(A) = P(A\Gamma)$$

for every orthogonal Γ and every Borel set A. See Anderson (1958), Chapter 13. General theory of Haar measures on topological groups can be found in

Nachbin (1965) or Halmos (1974), for example. For the group of orthogonal matrices, the existence and uniqueness of the Haar invariant distribution can be established easily. For the uniqueness we have

Lemma 1. *Let two probability measures P_1, P_2 satisfy (1). Then $P_1(A) = P_2(A)$ for every Borel set A. Furthermore $P_1(A) = P_1(A')$ where $A' = \{ H' \mid H \in A \}$.*

Proof. Let H_1, H_2 be independently distributed according to P_1, P_2 respectively. Then

$$(2) \quad Pr(H_1 H_2' \in A) = \mathcal{E}_{H_2}\{Pr(H_1 H_2' \in A \mid H_2)\} = \mathcal{E}_{H_2}\{P_1(A)\} = P_1(A).$$

Similarly
$$(3) \quad Pr(H_1 H_2' \in A) = Pr(H_2 H_1' \in A') = \mathcal{E}_{H_1}\{Pr(H_2 H_1' \in A' \mid H_1)\} = P_2(A').$$

Hence

$$(4) \qquad\qquad P_1(A) = P_2(A').$$

Putting $P_1 = P_2$ we obtain $P_1(A) = P_1(A')$, $P_2(A) = P_2(A')$. Substituting this into (4) we obtain $P_1(A) = P_2(A)$. ∎

Remark 1. For a more rigorous proof (2) and (3) have to be converted to the form of Fubini's theorem, as is done in standard proofs (see Section 60 of Halmos (1974)). The same remark applies to the proof of Lemma 3 below. Also note that the second assertion of Lemma 1 shows that if H is uniform then H' is uniform.

Existence can be very explicitly established as follows.

Lemma 2. *Let $U = (u_{ij})$ be a $k \times k$ matrix such that u_{ij} are independent standard normal variables. Then with probability 1, U can be uniquely expressed as*

$$(5) \qquad\qquad U = TH,$$

where $T = (t_{ij})$ is lower triangular with positive diagonal elements and H is orthogonal. Furthermore (i) T and H are independent, (ii) H is uniform, (iii) t_{ij} are all independent and $t_{ii} \sim \chi(k-i+1)$, $t_{ij} \sim \mathcal{N}(0,1)$, $i > j$.

Proof. U is nonsingular with probability 1. Therefore suppose $|U| \neq 0$. Now performing the Gram-Schmidt orthonormalization to the rows of U starting from the first row we obtain $SU = H$ where S is lower triangular with positive diagonal elements and H is orthogonal. Letting $T = S^{-1}$ we obtain (5). Since (5) corresponds to the uniquely defined Gram-Schmidt orthonormalization T, H are unique. Now $W = UU' = TT'$ is distributed according to $\mathcal{W}(I_k, k)$. Hence (iii) follows from Lemma 3.1.3. To show (i) and (ii) we first note that for any orthogonal Γ, $U\Gamma$ has the same distribution as U. Furthermore $U\Gamma = T(H\Gamma)$. Therefore $H\Gamma$ is the resulting orthogonal matrix obtained by performing Gram-Schmidt orthonormalization to the rows of $U\Gamma$ and T is common to U and $U\Gamma$. This implies that given T the conditional distributions of H and $H\Gamma$ are the same. Therefore the conditional distribution of H given T is uniform. Now by unconditioning we see that T and H are independent and H has the uniform distribution. ∎

This lemma has been known for a long time. See Kshirsagar (1959), Example 6 in Chapter 8 of Lehmann (1959), Saw (1970) for example.

Now we prove the following fundamental identity (James (1961a)). The proof is a modification of one in Saw (1977).

Theorem 1. *Let A, B be $k \times k$ symmetric matrices. Then*

(6) $$\mathcal{E}_H \mathcal{Y}_p(AHBH') = \mathcal{Y}_p(A)\mathcal{Y}_p(B)/\mathcal{Y}_p(I_k),$$

where $k \times k$ orthogonal H has the uniform distribution.

Proof. Let $f(A, B)$ denote the left hand side of (6). Let $\lambda(A) = \alpha = (\alpha_1, \ldots, \alpha_k)$ and $\lambda(B) = \beta = (\beta_1, \ldots, \beta_k)$. Let $A = H_1 D_1 H_1'$, $B = H_2 D_2 H_2'$, where H_1, H_2 are orthogonal and $D_1 = \text{diag}(\alpha_1, \ldots, \alpha_k)$, $D_2 = \text{diag}(\beta_1, \ldots, \beta_k)$. Now

(7) $$\begin{aligned}\mathcal{Y}_p(AHBH') &= \mathcal{Y}_p(H_1 D_1 H_1' H H_2 D_2 H_2' H') \\ &= \mathcal{Y}_p(D_1 H_3 D_2 H_3'),\end{aligned}$$

where $H_3 = H_1'HH_2$ which has the uniform distribution. Therefore

(8) $$f(A, B) = \mathcal{E}_H \mathcal{Y}_p(D_1 H D_2 H').$$

This depends only on $\alpha = (\alpha_1, \ldots, \alpha_k)$, and $\beta = (\beta_1, \ldots, \beta_k)$. Now for any permutation matrix P, $A = (H_1 P)(P' D_1 P)(P' H_1')$. Noting that a permutation matrix is orthogonal we get $\mathcal{E}_H \mathcal{Y}_p(D_1 H D_2 H') = \mathcal{E}_H \mathcal{Y}_p(P' D_1 P H D_2 H')$. therefore $f(A, B)$ is symmetric in $\alpha_1, \ldots, \alpha_k$. Similarly it is symmetric in $\beta_1, \ldots, \beta_k$. Now on the left hand side of (6) express $\mathcal{Y}_p$ in terms of $\mathcal{U}_q$'s. Furthermore for each elementary symmetric function $u_r = u_r(AHBH')$ constituting $\mathcal{U}_q$ use the relation (3.1.4). We see that $\mathcal{Y}_p(AHBH')$ and hence $f(A, B)$ are polynomials in $(\alpha_1, \ldots, \alpha_k, \beta_1, \ldots, \beta_k)$. Suppose that $f(A, B)$ is completely expanded into monomial terms. Consider the term of the form $c\alpha_1^{q_1} \cdots \alpha_\ell^{q_\ell}, (q = (q_1, \ldots, q_\ell) \in \mathcal{P}_n)$ in $f(A, B)$. By symmetry among α_i's $f(A, B)$ has the term $c\alpha_{i_1}^{q_1} \cdots \alpha_{i_\ell}^{q_\ell}$ with the same coefficient c. Collecting these permuted terms in α's we obtain $c M_q(\alpha)$. However c is a polynomial in β's and by symmetry among β's c can be written as a linear combination of $M_q(\beta)$'s. Collecting all terms we can write

$$f(A, B) = \sum_{q,q'} a_{qq'} M_q(\alpha) M_{q'}(\beta)$$

for some real numbers $a_{qq'}$. Expressing M_q's in terms of $\mathcal{Y}_q$'s we alternatively have

$$f(A, B) = \sum_{q,q'} c_{qq'} \mathcal{Y}_q(A) \mathcal{Y}_{q'}(B).$$

Note that $c_{qq'} = c_{q'q}$ because $\mathcal{Y}_p(AHBH') = \mathcal{Y}_p(BH'AH)$ and H' has the uniform distribution (see Remark 1). Now let A be distributed independently of H and according to $\mathcal{W}(\Sigma, \nu_0)$ where ν_0 is such that $\lambda_{\nu_0 p}, p \in \mathcal{P}$ are all different (see the proof of Lemma 3.1.4). Then by Corollary 3.1.3

(9) $$\mathcal{E}_A \mathcal{E}_H \mathcal{Y}_p(AHBH') = \sum_{q,q'} c_{qq'} \lambda_{\nu_0 q} \mathcal{Y}_q(\Sigma) \mathcal{Y}_{q'}(B).$$

On the other hand taking expectation with respect to A first we obtain

(10) $$\mathcal{E}_H \mathcal{E}_A \mathcal{Y}_p(AHBH') = \lambda_{\nu_0 p} \mathcal{E}_H \mathcal{Y}_p(\Sigma H B H')$$
$$= \lambda_{\nu_0 p} \sum_{q,q'} c_{qq'} \mathcal{Y}_q(\Sigma) \mathcal{Y}_{q'}(B).$$

Therefore

(11) $$0 = \sum_{q,q'}(\lambda_{\nu_0 p} - \lambda_{\nu_0 q})c_{qq'}\mathcal{Y}_q(\pmb{\Sigma})\mathcal{Y}_{q'}(\pmb{B}).$$

This holds for any $\pmb{\Sigma}$ and $\pmb{B}$. Hence $(\lambda_{\nu_0 p} - \lambda_{\nu_0 q})c_{qq'} = 0$ for all q,q'. Since $\lambda_{\nu_0 p} \neq \lambda_{\nu_0 q}$ for $p \neq q$ we have $c_{qq'} = 0$ for all q' and all $q \neq p$. But $c_{qq'} = c_{q'q}$. Therefore $c_{qq'} = 0$ unless $q = q' = p$. Therefore

(12) $$\mathcal{E}_H \mathcal{Y}_p(\pmb{A}\pmb{H}\pmb{B}\pmb{H'}) = c_{pp}\mathcal{Y}_p(\pmb{A})\mathcal{Y}_p(\pmb{B}).$$

Putting $\pmb{B} = \pmb{I}_k$ we obtain

(13) $$\mathcal{Y}_p(\pmb{A}) = c_{pp}\mathcal{Y}_p(\pmb{I}_k)\mathcal{Y}_p(\pmb{A}).$$

Hence $c_{pp}\mathcal{Y}_p(\pmb{I}_k) = 1$ and this proves the theorem. ∎

For more about this proof see Section 4.1.

Theorem 1 implies the following rather strong result.

Theorem 2. *Suppose that a $k \times k$ random symmetric matrix $\pmb{V}$ has a distribution such that for every orthogonal $\pmb{\Gamma}$, $\pmb{\Gamma}\pmb{V}\pmb{\Gamma'}$ has the same distribution as $\pmb{V}$. Then for symmetric $\pmb{A}$*

(14) $$\mathcal{E}_V \mathcal{Y}_p(\pmb{A}\pmb{V}) = c_p \mathcal{Y}_p(\pmb{A}),$$

where

(15) $$c_p = \mathcal{E}_V\{\mathcal{Y}_p(\pmb{V})\}/\mathcal{Y}_p(\pmb{I}_k).$$

Proof. As in the proof of Lemma 3.1.2 $\mathcal{E}_V \mathcal{Y}_p(\pmb{A}\pmb{V}) \in V_n$. Now since the distribution of $\pmb{\Gamma}\pmb{V}\pmb{\Gamma'}$ is the same as $\pmb{V}$ we have

(16) $$\mathcal{E}_V \mathcal{Y}_p(\pmb{A}\pmb{\Gamma}\pmb{V}\pmb{\Gamma'}) = \mathcal{E}_V \mathcal{Y}_p(\pmb{A}\pmb{V}).$$

Letting $\pmb{\Gamma}$ be uniformly distributed independently of $\pmb{V}$

(17) $$\begin{aligned}\mathcal{E}_V \mathcal{Y}_p(\pmb{A}\pmb{V}) &= \mathcal{E}_\Gamma \mathcal{E}_V \mathcal{Y}_p(\pmb{A}\pmb{\Gamma}\pmb{V}\pmb{\Gamma'}) \\ &= \mathcal{E}_V \mathcal{E}_\Gamma \mathcal{Y}_p(\pmb{A}\pmb{\Gamma}\pmb{V}\pmb{\Gamma'}) \\ &= \mathcal{E}_V\{\mathcal{Y}_p(\pmb{A})\mathcal{Y}_p(\pmb{V})/\mathcal{Y}_p(\pmb{I}_k)\} \\ &= \mathcal{Y}_p(\pmb{A})\mathcal{E}_V\{\mathcal{Y}_p(\pmb{V})\}/\mathcal{Y}_p(\pmb{I}_k).\end{aligned}$$

∎

Remark 2. In the sequel we call the distribution of V "orthogonally invariant" if it satisfies the condition of Theorem 2.

Although Theorem 2 has not been explicitly stated, it has been implicitly used for several cases; first with the multivariate beta distribution by Constantine (1963), later with the inverted Wishart distribution by Khatri (1966) etc. These cases will be examined in Section 4.3 together with the evaluation of c_p for each case.

We note that Theorem 2 is a generalization of Theorem 3.1.1. Now suppose that we chose V which has an orthogonally invariant distribution instead of the Wishart matrix W for the construction of zonal polynomials in Section 3.1. Then the construction could have been carried out in exactly the same way provided that c_p, $p \in P_n$ in Theorem 2 are all distinct for V. Furthermore if we examine the proof of Theorem 1 closely we find that we could take $A = \Sigma^{\frac{1}{2}} V \Sigma^{\frac{1}{2}}$ in (9) and (10). Once Theorem 1 is proved the identity involving the Wishart distribution can be derived as a special case. Although the Wishart distribution seems to be a natural candidate to take for our construction, we could have used any orthogonally invariant distribution from a purely logical point of view.

Orthogonally invariant distributions are characterized as follows.

Lemma 3. Let $V = HDH'$ where H is orthogonal and D is diagonal. Let H and D be independently distributed such that H has the uniform distribution. (Diagonal elements of D can have any distribution.) Then V has an orthogonally invariant distribution. Conversely all orthogonally invariant distributions can be obtained in this way.

Proof. The first part of the lemma is obvious. To prove the converse suppose that V has an orthogonally invariant distribution. Now we form a new random matrix $\tilde{V} = HVH'$ where H has the uniform distribution independently of V. Then $\tilde{V}$ has the same distribution as V because for any Borel set A

$$
(18) \quad \begin{aligned} Pr(\tilde{V} \in A) &= \mathcal{E}_H\{Pr(HVH' \in A \mid H)\} \\ &= \mathcal{E}_H\{Pr(V \in A)\} = Pr(V \in A). \end{aligned}
$$

§ 3.2   *Integral identities*   31

Now we evaluate $Pr(\tilde{V} \in A)$ by conditioning on V. For fixed V we can write $V = \Gamma D \Gamma'$ where Γ is orthogonal and $D = \text{diag}(d_1, \ldots, d_k)$. We require $d_1 \geq \cdots \geq d_k$ then $D = D(V)$ is unique. Then

$$
(19) \quad \begin{aligned} Pr(\tilde{V} \in A \mid V) &= Pr(H\Gamma D(V)\Gamma' H' \in A \mid V) \\ &= Pr(HD(V)H' \in A \mid V). \end{aligned}
$$

Note that we replaced $H\Gamma$ by H since $H\Gamma$ has the uniform distribution. Hence

$$
(20) \quad \begin{aligned} Pr(\tilde{V} \in A) &= \mathcal{E}_V\{Pr(\tilde{V} \in A \mid V)\} \\ &= \mathcal{E}_V\{Pr(HD(V)H' \in A \mid V)\} \\ &= Pr(HD(V)H' \in A). \end{aligned}
$$

This proves the lemma. ∎

Remark 3. Note that the set of orthogonally invariant distributions is convex with respect to taking mixture of distributions. Lemma 3 implies that the extreme points of this convex set are given by those distributions for which D is degenerate.

We can replace H in Theorem 1 by U whose elements are independent normal variables.

Theorem 3. *Let $U = (u_{ij})$ be a $k \times k$ matrix such that u_{ij} are independent standard normal variables. Then for symmetric A, B*

$$
(21) \quad \mathcal{E}_U \mathcal{Y}_p(AUBU') = \frac{\lambda_{kp}}{\mathcal{Y}_p(I_k)} \mathcal{Y}_p(A) \mathcal{Y}_p(B).
$$

Proof. By Lemma 2 $U = TH$. Then

$$
(22) \quad \begin{aligned} \mathcal{E}_U \mathcal{Y}_p(AUBU') &= \mathcal{E}_T \mathcal{E}_H \mathcal{Y}_p(ATHBH'T') \\ &= \mathcal{E}_T \mathcal{E}_H \mathcal{Y}_p(T'ATHBH') \\ &= \mathcal{E}_T \mathcal{Y}_p(T'AT) \mathcal{Y}_p(B)/\mathcal{Y}_p(I_k) \\ &= \mathcal{E}_T \mathcal{Y}_p(ATT') \mathcal{Y}_p(B)/\mathcal{Y}_p(I_k) \\ &= \frac{\lambda_{kp}}{\mathcal{Y}_p(I_k)} \mathcal{Y}_p(A) \mathcal{Y}_p(B). \end{aligned}
$$

We used the fact that $TT' = UU' \sim \mathcal{W}(I_k, k)$. ∎

Theorem 3 leads to the following important observation.

Theorem 4. $b_p \equiv \lambda_{kp}/\mathcal{Y}_p(I_k)$ *is a constant independent of k.*

Proof. Let A, B be augmented by zeros as

$$\tilde{A} = \begin{pmatrix} A & 0 \\ 0 & 0 \end{pmatrix}, \quad k_1 \times k_1, \qquad \tilde{B} = \begin{pmatrix} B & 0 \\ 0 & 0 \end{pmatrix}, \quad k_1 \times k_1.$$

Then $\mathcal{Y}_p(\tilde{A}) = \mathcal{Y}_p(A)$, $\mathcal{Y}_p(\tilde{B}) = \mathcal{Y}_p(B)$, and $\mathcal{Y}_p(\tilde{A}\tilde{U}\tilde{B}\tilde{U}') = \mathcal{Y}_p(AUBU')$ where $\tilde{U}(k_1 \times k_1)$ is obtained by adding independent standard normal variables to U. Now (21) implies the result. ∎

We evaluate the b_p's for a particular normalization of zonal polynomial to be denoted as $_1\mathcal{Y}_p$ in Section 4.2. Corresponding to Theorem 2, Theorem 3 can be generalized as follows.

Theorem 5. *Let X be a $k \times k$ random matrix (not necessarily symmetric) such that for every orthogonal Γ_1, Γ_2, the distribution of $\Gamma_1 X \Gamma_2$ is the same as the distribution of X. Then for symmetric A, B*

(23) $$\mathcal{E}_X \mathcal{Y}_p(AXBX') = \gamma_p \mathcal{Y}_p(A) \mathcal{Y}_p(B),$$

where

(24) $$\gamma_p = \mathcal{E}_X\{\mathcal{Y}_p(XX')\}/\{\mathcal{Y}_p(I_k)\}^2.$$

Proof. For any orthogonal Γ_1

(25) $$\mathcal{E}_X \mathcal{Y}_p(AXBX') = \mathcal{E}_X \mathcal{Y}_p(AX\Gamma_1 B\Gamma_1' X').$$

Letting Γ_1 be uniformly distributed we obtain

(26) $$\mathcal{E}_X \mathcal{Y}_p(AXBX') = \frac{\mathcal{Y}_p(B)}{\mathcal{Y}_p(I_k)} \mathcal{E}_X \mathcal{Y}_p(X'AX)$$
$$= \frac{\mathcal{Y}_p(B)}{\mathcal{Y}_p(I_k)} \mathcal{E}_X \mathcal{Y}_p(AXX').$$

Now $V = XX'$ has an orthogonally invariant distribution because $\Gamma_2 V \Gamma_2' = (\Gamma_2 X)(\Gamma_2 X)'$. Therefore by Theorem 2

(27) $$\mathcal{E}_X \mathcal{Y}_p(AXX') = \mathcal{Y}_p(A) \mathcal{E}_X\{\mathcal{Y}_p(XX')\}/\mathcal{Y}_p(I_k).$$

Substituting (27) into (26) we obtain the theorem. ∎

Remark 4. We call the distribution of X "orthogonally biinvariant" if it satisfies the condition of Theorem 5.

Corresponding to Lemma 3 we have

Lemma 4. *Let $X = H_1 D H_2$ where H_1, H_2 are orthogonal and D is diagonal. Let H_1, H_2, D be independently distributed such that H_1, H_2 have the uniform distribution. (D can have any distribution.) Then X has an orthogonally biinvariant distribution. Conversely all orthogonally biinvariant distributions can be obtained in this way.*

The proof is entirely analogous to the proof of Lemma 3, therefore we omit it.

Remark 5. The notion of orthogonal biinvariance can be applied to rectangular matrices. If X is $k \times m$ in Theorem 5 we obtain

$$(28) \qquad \gamma_p = \frac{\mathcal{E}_X \mathcal{Y}_p(XX')}{\mathcal{Y}_p(I_k)\mathcal{Y}_p(I_m)}$$

and in Lemma 4 (for $k \leq m$) we replace $X = H_1 D H_2$ by $X = H_1(D, 0)H_2$.

In the sequel we almost exclusively work with the Wishart and the normal distributions. But in view of Theorem 2 and Theorem 5 there could be other distributions which give information on various aspects of zonal polynomials.

§ 3.3 AN INTEGRAL REPRESENTATION OF ZONAL POLYNOMIALS

We prove an integral representation by Kates (1980) which shows that (i) zonal polynomials are positive for positive definite A and increasing in each root of A, (ii) in the normalization Z_p defined below the coefficients a_{pq} in $Z_p = \sum a_{pq} M_q$ are nonnegative integers. The derivation by Kates is rather abstract but the integral representation can be proved directly in our framework. The representation can be formulated in several ways. James (1973) derived one involving uniform orthogonal matrix. We discuss these variations in Section 4.7.

From Theorem 3.2.4 we see that a constant b_p or equivalently the value of a zonal polynomial at I_k describes a particular normalization. The normalization Z_p is the simplest one in this sense.

Definition 1. A particular normalization of a zonal polynomial denoted by Z_p is defined by

$$Z_p(I_k) = \lambda_{kp}, \tag{1}$$

or $b_p = 1$ in Theorem 3.2.4.

Theorem 1. *(Kates, 1980)* Let $p = (p_1, \ldots, p_\ell)$. For $k \times k$ symmetric A

$$Z_p(A) = \mathcal{E}_U\{\Delta_1^{p_1-p_2} \Delta_2^{p_2-p_3} \cdots \Delta_\ell^{p_\ell}\}, \tag{2}$$

where $\Delta_i = UAU'(1, \ldots, i)$ is the determinant of the $i \times i$ upper left minor of UAU' and U is a $k \times k$ random matrix whose entries are independent standard normal variables.

Proof. For symmetric A let

$$f(A) = \mathcal{E}_U\{\Delta_1^{p_1-p_2} \cdots \Delta_\ell^{p_\ell}\}. \tag{3}$$

It can be routinely checked that f is a homogeneous symmetric polynomial of degree $n = |p|$ in the roots of A. Furthermore augmenting A to $\tilde{A}$ ($k_1 \times k_1$) by adding zeros and augmenting U to $\tilde{U}$ by adding independent standard normal variables do not change the upper left part of UAU'. Therefore (3) does not depend on k. Hence $f \in V_n$. Note that we can extend the definition of f to nonsymmetric matrices as well (see Remark 3.1.6). Now we want to show

$$(\tau_\nu f)(A) = \lambda_{\nu p} f(A) \tag{4}$$

for all sufficiently large ν and for all symmetric A. Let

$$\tilde{A} = \begin{pmatrix} A & 0 \\ 0 & 0 \end{pmatrix} \quad : \nu \times \nu$$

§ 3.3 *An integral representation* 35

and $\tilde{W} = Y'Y$ where Y is a $\nu \times \nu$ matrix whose entries are standard normal variables. Then

(5)
$$(\tau_\nu f)(A) = \mathcal{E}_{\tilde{W}}\{f(\tilde{A}\tilde{W})\} = \mathcal{E}_Y\{f(Y\tilde{A}Y')\}$$
$$= \mathcal{E}_Y \mathcal{E}_{\tilde{U}}\{\prod_{i=1}^{\ell} [\tilde{U}Y\tilde{A}Y'\tilde{U}'(1,\ldots,i)]^{p_i - p_{i+1}}\}$$
$$= \mathcal{E}_{\tilde{U}} \mathcal{E}_Y\{\prod_{i=1}^{\ell} [Y\tilde{U}\tilde{A}\tilde{U}'Y'(1,\ldots,i)]^{p_i - p_{i+1}}\}.$$

We switched $\tilde{U}$ and Y because they have the same distribution. Now by Lemma 3.2.2 $Y = TH$ and H can be absorbed into U. Therefore

(6)
$$\mathcal{E}_{\tilde{U}} \mathcal{E}_Y\{\prod_{i=1}^{\ell} [Y\tilde{U}\tilde{A}\tilde{U}'Y'(1,\ldots,i)]^{p_i - p_{i+1}}\}$$
$$= \mathcal{E}_{\tilde{U}} \mathcal{E}_T\{\prod_{i=1}^{\ell} [T\tilde{U}\tilde{A}\tilde{U}'T'(1,\ldots,i)]^{p_i - p_{i+1}}\}$$
$$= \mathcal{E}_{\tilde{U}} \mathcal{E}_T\{\prod_{i=1}^{\ell} (t_{11}^2 \cdots t_{ii}^2)^{p_i - p_{i+1}} [\tilde{U}\tilde{A}\tilde{U}'(1,\ldots,i)]^{p_i - p_{i+1}}\}$$
$$= \lambda_{\nu p} \mathcal{E}_{\tilde{U}}\{\prod_{i=1}^{\ell} [\tilde{U}\tilde{A}\tilde{U}'(1,\ldots,i)]^{p_i - p_{i+1}}\}$$
$$= \lambda_{\nu p} f(A).$$

Hence $f = \mathcal{Y}_p$ by Theorem 3.1.1. Putting $A = I_k$ we obtain

(7)
$$f(I_k) = \mathcal{E}_W\{W(1)^{p_1 - p_2} \cdots W(1,\ldots,\ell)^{p_\ell}\},$$

where $W \sim \mathcal{W}(I_k, k)$. Again by the triangular decomposition $W = TT'$ (Lemma 3.1.3) we obtain $f(I_k) = \lambda_{kp}$. Therefore $f = Z_p$. ∎

Note that the coefficients of the monomial terms in Z_p are integers, being the expected value of sum of products of independent standard normal variables. Furthermore if $A = \text{diag}(\alpha_1, \ldots, \alpha_k)$ then by the Binet-Cauchy theorem (see Gantmacher (1959) for example)

$$
\begin{aligned}
\boldsymbol{UAU}'(1,\ldots,r) &= \sum_{i_1<\cdots<i_r}\sum_{j_1<\cdots<j_r} U\begin{pmatrix}1,\ldots,r\\i_1,\ldots,i_r\end{pmatrix} A\begin{pmatrix}i_1,\ldots,i_r\\j_1,\ldots,j_r\end{pmatrix} U'\begin{pmatrix}j_1,\ldots,j_r\\1,\ldots,r\end{pmatrix} \\
&= \sum_{i_1<\cdots<i_r} \alpha_{i_1}\cdots\alpha_{i_r}\left\{U\begin{pmatrix}1,\ldots,r\\i_1,\ldots,i_r\end{pmatrix}\right\}^2,
\end{aligned}
$$
(8)

where $\boldsymbol{B}\begin{pmatrix}i_1,\ldots,i_r\\j_1,\ldots,j_r\end{pmatrix}$ denotes the determinant of a minor formed by rows $i_1,\ldots,i_r$ and columns $j_1,\ldots,j_r$ of $\boldsymbol{B}$. (8) is obviously increasing in each α_i when $\boldsymbol{A}$ is positive definite. Furthermore coefficients for monomial terms are nonnegative. These points are discussed in Kates (1980). For more about this see Section 4.1. Generalizations of Theorem 1 will be discussed in Section 4.7.

§ 3.4 A GENERATING FUNCTION OF ZONAL POLYNOMIALS

One of the main contributions of Saw (1977) is his generating function which gives a relatively simple way of computing zonal polynomials. Let

(1) $$(\operatorname{tr}\boldsymbol{C})^n = \mathcal{U}_p(\boldsymbol{C}) = \sum_{p\in\mathcal{P}_n} d_p Z_p(\boldsymbol{C}).$$

Let $\boldsymbol{C} = \boldsymbol{AUBU}'$ where $\boldsymbol{A} = \operatorname{diag}(\alpha_1,\ldots,\alpha_k)$, $\boldsymbol{B} = \operatorname{diag}(\beta_1,\ldots,\beta_k)$ and the elements of $\boldsymbol{U}$ are independent standard normal variables. Then by Theorem 3.2.3

$$
\begin{aligned}
\mathcal{E}_U(\operatorname{tr}\boldsymbol{AUBU}')^n &= \sum_{p\in\mathcal{P}_n} d_p \mathcal{E}_U Z_p(\boldsymbol{AUBU}') \\
&= \sum_{p\in\mathcal{P}_n} d_p \frac{\lambda_{kp}}{Z_p(\boldsymbol{I}_k)} Z_p(\boldsymbol{A}) Z_p(\boldsymbol{B}) \\
&= \sum_{p\in\mathcal{P}_n} d_p Z_p(\boldsymbol{A}) Z_p(\boldsymbol{B}).
\end{aligned}
$$
(2)

Therefore for sufficiently small θ

$$
\begin{aligned}
\mathcal{E}_U\{\exp(\theta \operatorname{tr} \boldsymbol{AUBU}')\} \\
= \mathcal{E}_U\{\sum_{n=0}^{\infty}(\theta^n/n!)(\operatorname{tr} \boldsymbol{AUBU}')^n\} \\
= \sum_{n=0}^{\infty}(\theta^n/n!)\sum_{p\in P_n} d_p Z_p(\boldsymbol{A})Z_p(\boldsymbol{B}).
\end{aligned}
\tag{3}
$$

On the other hand

$$
\operatorname{tr} \boldsymbol{AUBU}' = \sum_{i,j}^{k} \alpha_i \beta_j u_{ij}^2.
\tag{4}
$$

Hence for sufficiently small θ

$$
\begin{aligned}
\mathcal{E}_U\{\exp(\theta \operatorname{tr} \boldsymbol{AUBU}')\} &= \mathcal{E}_U\{\exp(\theta \sum_{i,j}^{k} \alpha_i \beta_j u_{ij}^2)\} \\
&= \prod_{i,j}^{k}(1 - 2\theta\alpha_i\beta_j)^{-\frac{1}{2}}.
\end{aligned}
\tag{5}
$$

From (3) and (5) we obtain

Theorem 1.

$$
\prod_{i,j}^{k}(1 - 2\theta\alpha_i\beta_j)^{-\frac{1}{2}} = \sum_{n=0}^{\infty}(\theta^n/n!)\sum_{p\in P_n} d_p Z_p(\boldsymbol{A})Z_p(\boldsymbol{B}).
\tag{6}
$$

The left hand side of (6) can be expanded as follows.

$$\prod_{i,j}^{k}(1 - 2\theta\alpha_i\beta_j)^{-\frac{1}{2}}$$

$$= \exp\{\log \prod_{i,j}^{k}(1 - 2\theta\alpha_i\beta_j)^{-\frac{1}{2}}\}$$

$$= \exp\{\frac{1}{2}\sum_{i,j}^{k}\sum_{r=1}^{\infty}\frac{(2\theta)^r}{r}\alpha_i^r\beta_j^r\}$$

(7) $$= \exp\{\frac{1}{2}\sum_{r=1}^{\infty}\frac{(2\theta)^r}{r}t_r(A)t_r(B)\}$$

$$= \sum_{n=0}^{\infty}\sum_{p\in P_n}(2\theta)^n T_p(A)T_p(B)$$

$$\cdot \frac{1}{p_1!}\left(\frac{1}{2}\right)^{p_1}\binom{p_1}{p_1-p_2, p_2-p_3, \ldots, p_{\ell(p)}}\left(\prod_{r=1}^{\ell(p)}r^{p_r-p_{r+1}}\right)^{-1}$$

$$= \sum_{n=0}^{\infty}(\theta^n/n!)\sum_{p\in P_n}c_p T_p(A)T_p(B),$$

where

(8) $$c_p = |p|!2^{|p|-h(p)}\left\{\prod_{r=1}^{\ell(p)}r^{p_r-p_{r+1}}(p_r-p_{r+1})!\right\}^{-1}.$$

The fourth equality follows from the fact that T_p being a product of p_1 terms comes only from the p_1-*th* power term in the expansion of exp. Comparing the coefficient of θ^n in (6) and (7) we obtain

(9) $$\sum_{p\in P_n}d_p Z_p(A)Z_p(B) = \sum_{p\in P_n}c_p T_p(A)T_p(B).$$

Note that c_p is positive for every $p \in P_n$. Hence the right hand side of (9) is a positive definite quadratic form. Now the left hand side of (9) is the same positive definite quadratic form expressed with the different basis $\{Z_p\}$. The positive definiteness implies $d_p > 0$ for every $p \in P_n$. Now let $D = \text{diag}(d_p, p \in$

§ 3.4   A generating function   39

$\mathcal{P}_n$), $C = \text{diag}(c_p, p \in \mathcal{P}_n)$. We recall that $Z = \mathcal{B}\mathcal{U}$ where $\mathcal{B}$ is upper triangular and $\mathcal{T} = F\mathcal{U}$ where F is lower triangular (see(2.2.28)). Therefore in matrix notation (9) is written as

(10) $$\mathcal{U}(A)'\mathcal{B}'D\mathcal{B}\mathcal{U}(B) = \mathcal{U}(A)'F'CF\mathcal{U}(B),$$

or

(11) $$\mathcal{B}'D\mathcal{B} = F'CF.$$

We note that the left hand side and the right hand side correspond to two different triangular decompositions of the same symmetric positive definite matrix. F can be computed from (2.2.24) or alternatively F can be obtained from tables given in David, Kendall, and Barton (1966) for $n \leq 12$. Therefore we can compute the right hand side of (9) relatively easily, then we decompose the resulting positive definite matrix as $\mathcal{B}'D\mathcal{B}$. Diagonal elements of $\mathcal{B}$ corresponding to Z_p is obtained in (4.2.7). This determines D and $\mathcal{B}$ uniquely.

Remark 1.   In terms of $\mathcal{M}_p$'s (11) can be written as $A'\mathcal{B}'D\mathcal{B}A = A'F'CFA$ where A is given in (2.2.14). Saw (1977) defined zonal polynomials or the upper triangular coefficient matrix $\mathcal{B}A$ by this relation and derived the first part of Theorem 3.1.1 from this definition. It seems that (11) should be looked at as providing a convenient algorithm for obtaining $\mathcal{B}$ rather than providing a definition of zonal polynomials because it lacks the conceptual motivation necessary for a definition.

Actually d_p is known to be (James (1964), formula(18))

(12) $$\begin{aligned} d_p &= \chi_{[2p]}(1)2^n n!/(2n)! \\ &= \frac{2^n n! \prod_{i<j}(2p_i - 2p_j - i + j)}{\prod_{i=1}^{\ell(p)}(2p_i + \ell(p) - i)!}, \end{aligned}$$

where $n = |p|$ and $\chi_{[2p]}(1) = (2n)! \prod_{i<j}(2p_i - 2p_j - i + j) / \prod_{i=1}^{\ell(p)}(2p_i + \ell(p) - i)!$ is "the dimension of the representation $(2p) = (2p_1, \ldots, 2p_{\ell(p)})$ of the symmetric group on $2n$ symbols." This is one thing we were unable to

obtain by our elementary approach. It was obtained by James (1961) using group representation theory. We will discuss this point again in Section 4.2 and Section 5.4.

$d_p Z_p$ is usually denoted by C_p so that (1) can be written simply as

(13) $$(\operatorname{tr} A)^n = \sum_{p \in P_n} C_p(A).$$

This notation often makes it simpler to write down various noncentral densities. Our last item in this section is related to this point.

Lemma 1.

(14)
$$\mathcal{E}_H (\operatorname{tr} AH)^{2n} = \sum_{p \in P_n} \frac{2^n n! d_p}{(2n)! \lambda_{kp}} Z_p(AA')$$
$$= \sum_{p \in P_n} \frac{2^n n!}{(2n)! \lambda_{kp}} C_p(AA'),$$

where $k \times k$ orthogonal H is uniformly distributed.

Proof. Let the singular value decomposition be $A = \Gamma_1 D \Gamma_2$ where Γ_1, Γ_2 are orthogonal, $D = \operatorname{diag}(\delta_1, \ldots, \delta_k)$ and $\theta_i = \delta_i^2, i = 1, \ldots, k$ are the characteristic roots of AA'. Then $(\operatorname{tr} AH)^{2n} = (\operatorname{tr} \Gamma_1 D \Gamma_2 H)^{2n} = (\operatorname{tr} D \Gamma_2 H \Gamma_1)^{2n}$ and $\Gamma_2 H \Gamma_1$ has the same distribution as H. Therefore $\mathcal{E}_H(\operatorname{tr} AH)^{2n}$ is a $2n$-th degree homogeneous polynomial in $\delta_1, \ldots, \delta_k$. Furthermore the order of $\delta_1, \ldots, \delta_k$ and the sign for each δ_i are arbitrary in the singular value decomposition. It follows that $\mathcal{E}_H(\operatorname{tr} AH)^{2n}$ is a homogeneous symmetric polynomial of degree n in $(\theta_1, \ldots, \theta_k)$. Therefore we can write

(15) $$\mathcal{E}_H(\operatorname{tr} AH)^{2n} = \sum_{p \in P_n} a_p Z_p(AA'),$$

for some real numbers a_p. Now let $A = \operatorname{diag}(\alpha_1, \ldots, \alpha_k)$ and $U = (u_{ij})$ be as before. Then $\operatorname{tr} AU = \sum \alpha_i u_{ii}$ is distributed according to $\mathcal{N}(0, \sum \alpha_i^2)$. Hence

(16)
$$\mathcal{E}_U(\operatorname{tr} AU)^{2n} = (\sum \alpha_i^2)^n \cdot 1 \cdot 3 \cdots (2n-1)$$
$$= \frac{(2n)!}{2^n n!} (\operatorname{tr} AA')^n$$
$$= \sum_{p \in P_n} \frac{(2n)! d_p}{2^n n!} Z_p(AA').$$

On the other hand by Lemma 3.2.2 and (15)

$$\begin{aligned}
\mathcal{E}_U(\operatorname{tr} \boldsymbol{A}\boldsymbol{U})^{2n} &= \mathcal{E}_{T,H}(\operatorname{tr} \boldsymbol{A}\boldsymbol{T}\boldsymbol{H})^{2n} \\
&= \sum_{p \in \mathcal{P}_n} a_p \mathcal{E}_T Z_p(\boldsymbol{A}\boldsymbol{T}\boldsymbol{T}'\boldsymbol{A}') \\
&= \sum_{p \in \mathcal{P}_n} a_p \lambda_{kp} Z_p(\boldsymbol{A}\boldsymbol{A}').
\end{aligned}$$

(17)

Comparing (16) and (17) we obtain (14). ∎

CHAPTER 4

More properties of zonal polynomials

This chapter is a collection of results which are for the most part generalizations and refinements of the basic results given in Chapter 3. A particular emphasis is placed on the coefficients of zonal polynomials. In this respect this chapter contains new results and presumably covers almost all known results. On the other hand we do not survey various known identities involving zonal polynomials. For this purpose the reader is referred to an excellent survey paper by Subrahmaniam (1976). Actually in the discussion of the orthogonally invariant distributions we saw that zonal polynomials satisfy an infinite number of identities. It is a rather frustrating fact that although many identities for zonal polynomials are already known, explicit forms of zonal polynomials are not known.

§ 4.1 MAJORIZATION ORDERING

The proof of Theorem 3.2.1 which played an essential role for the subsequent development in Chapter 3 is not complete as it is. In (3.2.11) we argued that

(1) $$0 = \sum_{q,q'} (\lambda_{\nu_0 p} - \lambda_{\nu_0 q}) c_{qq'} \mathcal{Y}_q(\Sigma) \mathcal{Y}_{q'}(B),$$

for all symmetric B and all positive semidefinite Σ implies $(\lambda_{\nu_0 p} - \lambda_{\nu_0 q}) c_{qq'} = 0$. One objection may be that Σ is restricted to be positive semidefinite.

But this causes no trouble since (1) is a polynomial and a polynomial which is identically zero for nonnegative arguments has to be zero everywhere. A more serious question is that the dimensionality of Σ and B is fixed to be $k \times k$ which is the dimensionality of the uniform orthogonal matrix H. The same objection applies to the proof of Lemma 3.4.1.

What we have to consider is the space of n-th degree homogeneous symmetric polynomials in k variables, where k is fixed. We denote this space by $V_{n,k}$. Let $m < k$. If $f(x_1, \ldots, x_k) \in V_{n,k}$ then $f(x_1, \ldots, x_m, 0, \ldots, 0) \in V_{n,m}$. In this sense $V_{n,m}$ can be considered as a subspace of $V_{n,k}$. Now from the argument in Section 2.2 it follows that if $k \geq n$ then $\{ M_p, p \in P_n \}$ forms a basis of $V_{n,k}$. Therefore if $k \geq n$, $V_{n,k}$ are all isomorphic to V_n. However if $k < n$ then for p such that $\ell(p) > k$ M_p is identically 0. Therefore $\dim V_{n,k} < \dim V_n$. In order to proceed further we have to identify bases of $V_{n,k}$ for $k < n$.

For this purpose we now study homogeneous symmetric polynomials again from the viewpoint of majorization ordering. The following is a refinement of Lemma 2.2.2.

Lemma 1.

$$\text{(2)} \qquad U_p = M_p + \sum_{q \prec p, q \neq p} a_{pq} M_q,$$

$$\text{(3)} \qquad M_p = U_p + \sum_{q \prec p, q \neq p} a^{pq} U_q.$$

Proof. For $1 \leq r < k$ let $\alpha = x_1 = \cdots = x_r$. Then the degree of α in M_q is $q_1 + \cdots + q_r$. The degree of α in U_p is $p_1 + \cdots + p_r$ because

$$\text{(4)} \qquad \begin{aligned} & U_p(\alpha, \ldots, \alpha, x_{r+1}, \ldots) \\ &= (\alpha + \cdots)^{p_1 - p_2}(\alpha^2 + \cdots)^{p_2 - p_3} \cdots \\ &\quad \cdot (\alpha^r + \cdots)^{p_r - p_{r+1}}(\alpha^r x_{r+1} \cdots)^{p_{r+1} - p_{r+2}} \cdots (\alpha^r x_{r+1} \cdots x_\ell + \cdots)^{p_\ell} \\ &= c\alpha^{(p_1 - p_2) + 2(p_2 - p_3) + \cdots + r(p_r - p_{r+1}) + \cdots + rp_\ell} + \cdots \\ &= c\alpha^{p_1 + \cdots + p_r} + \cdots, \end{aligned}$$

where c is a term not containing α. Let $Q_r = \{q \mid q < p \text{ and } q_1 + \cdots + q_r > p_1 + \cdots + p_r\}$. Now since the degree of α in (2) is $p_1 + \cdots + p_r$ we have

(5) $$\sum_{q \in Q_r} a_{pq} M_q(\alpha, \ldots, \alpha, x_{r+1}, \ldots) = 0.$$

Now $M_q(\alpha, \ldots, \alpha, x_{r+1}, \ldots, x_k)$ are linearly independent if k is sufficiently large. Therefore $a_{pq} = 0$ for $q \in Q_r$. Repeating this argument for $r = 1, 2, \ldots$ we have

(6) $$a_{pq} = 0 \quad if \quad q \in Q_1 \cup Q_2 \cup \cdots.$$

But if q is not majorized by p there exists an r such that $q \in Q_r$. Therefore $a_{pq} = 0$ for every q which is not majorized by p. This proves (2). (3) can be proved similarly. ∎

Lemma 2. $\{M_p, p \in P_n, \ell(p) \leq k\}, \{U_p, p \in P_n, \ell(p) \leq k\}$ are bases of $V_{n,k}$.

Proof. Note that $M_p(x_1, \ldots, x_k) = 0$, $U_p(x_1, \ldots, x_k) = 0$ for p such that $\ell(p) > k$. Let $f \in V_{n,k}$. Then from (2.2.2)

(7) $$f(x_1, \ldots, x_k) = \sum_{p \in P_n} a_p M_p(x_1, \ldots, x_k)$$
$$= \sum_{p \in P_n, \ell(p) \leq k} a_p M_p(x_1, \ldots, x_k).$$

Therefore any $f \in V_{n,k}$ can be written as a linear combination of M_p's for which $p \in P_n, \ell(p) \leq k$. Now suppose

(8) $$\sum_{q \in P_n, \ell(q) \leq k} a_q M_q(x_1, \ldots, x_k) = 0.$$

Then differentiating (8) p_i times with respect to x_i, $i = 1, \ldots, \ell(p)$, (note $\ell(p) \leq k$) we obtain $a_p = 0$. Therefore $\{M_p, p \in P_n, \ell(p) \leq k\}$ is linearly independent in $V_{n,k}$. This shows that $\{M_p, p \in \mathcal{P}_n, \ell(p) \leq k\}$ is a basis of $V_{n,k}$. To show that $\{\mathcal{U}_p, p \in \mathcal{P}_n, \ell(p) \leq k\}$ is a basis it suffices to observe

$$
\begin{aligned}
\mathcal{M}_p(x_1, \ldots, x_k) &= \mathcal{U}_p(x_1, \ldots, x_k) + \sum_{q<p} a^{pq} \mathcal{U}_q(x_1, \ldots, x_k) \\
&= \mathcal{U}_p(x_1, \ldots, x_k) + \sum_{q<p, \ell(q) \leq k} a^{pq} \mathcal{U}_q(x_1, \ldots, x_k).
\end{aligned}
\tag{9}
$$

This and (7) with f replaced by $\mathcal{U}_p$ shows that $\{\mathcal{U}_p, p \in \mathcal{P}_n, \ell(p) \leq k\}$ is another basis of $V_{n,k}$. ∎

Remark 1. It is known that a_{pq} in (2) is nonzero and positive if and only if $q \prec p$. This is called the Gale-Ryser theorem. (See Macdonald (1979), Marshall and Olkin (1979).)

Now we prove the following.

Theorem 1.

$$
y_p = \sum_{q \prec p} a_{pq} \mathcal{U}_q = \sum_{q \prec p} a'_{pq} \mathcal{M}_q
\tag{10}
$$

for some real numbers a_{pq}, a'_{pq} and $\{y_p, p \in \mathcal{P}_n, \ell(p) \leq k\}$ forms a basis of $V_{n,k}$.

Proof. We first note that majorization is transitive, i.e. if $p^1 \succ p^2, p^2 \succ p^3$ then $p^1 \succ p^3$. Therefore in view of Lemma 1 the equalities involving $\mathcal{U}$'s and $\mathcal{M}$'s are equivalent. Hence we prove one involving $\mathcal{U}$'s. Now as in the proof of (2), the right hand side of (3.1.10)

$$
\mathcal{E}_W (\sum_{i_1} \alpha_{i_1} W(i_1))^{p_1 - p_2} (\sum_{i_1 < i_2} \alpha_{i_1} \alpha_{i_2} W(i_1, i_2))^{p_2 - p_3} \cdots
\tag{11}
$$

has only those monomial terms $\mathcal{M}_q(A)$ for which $q \prec p$. Therefore we can write

$$
\tau_\nu(\mathcal{U}_p) = \sum_{q \prec p} b_{pq} \mathcal{M}_q.
\tag{12}
$$

Substituting (3) into (12) and using the transitivity of majorization we obtain

(13) $$\tau_\nu(\mathcal{U}_p) = \sum_{q \prec p} b'_{pq} \mathcal{U}_q.$$

Now let

(14) $$y_p = \sum_{q \leq p} a_{pq} \mathcal{U}_q.$$

We want to show that $Q_p = \{q \mid a_{pq} \neq 0, q \text{ not majorized by } p\}$ is empty. We argue by contradiction. Suppose that Q_p is nonempty. Let q^* be the highest partition in Q_p. Then $a_{pq} \neq 0$ and $q > q^*$ imply $q \prec p$. For any such q

(15) $$a_{pq}\tau_\nu(\mathcal{U}_q) = a_{pq}\{\sum_{q' \prec q} b'_{qq'} \mathcal{U}_{q'}\}.$$

Now $q' \prec q$, $q \prec p$ imply $q' \prec p$. Hence the right hand side does not have $\mathcal{U}_{q^*}$ term. It follows that $\mathcal{U}_{q^*}$ does not appear in

(16) $$\sum_{q^* < q \leq p} a_{pq}\tau_\nu(\mathcal{U}_q).$$

Obviously

(17) $$\sum_{q < q^*} a_{pq}\tau_\nu(\mathcal{U}_q)$$

does not involve $\mathcal{U}_{q^*}$ term either. Therefore the coefficient of $\mathcal{U}_{q^*}$ in

(18) $$\begin{aligned}\tau_\nu(y_p) &= \sum_{q \leq p} a_{pq}\tau_\nu(\mathcal{U}_q) \\ &= a_{pq^*}\tau_\nu(\mathcal{U}_{q^*}) + (16) + (17)\end{aligned}$$

is $a_{pq^*}\lambda_{\nu q^*}$. On the other hand

(19) $$\tau_\nu(y_p) = \lambda_{\nu p} y_p.$$

§ 4.1   Majorization ordering   47

Therefore the coefficient of $\mathcal{U}_{q^*}$ on the right hand side of (19) is $\lambda_{\nu p} a_{pq^*}$. Taking $\nu = \nu_0$ we have a contradiction (see the proof of Lemma 3.1.4 for ν_0). Therefore Q_p is empty. This proves (10).

To prove the second assertion we note that $q \prec p$ implies $\ell(q) \geq \ell(p)$. Otherwise $p_1 + \cdots + p_{\ell(q)} < p_1 + \cdots + p_{\ell(p)} = n = q_1 + \cdots + q_{\ell(q)}$ and this contradicts $q \prec p$. Therefore in (10) we have only those $\mathcal{U}_q$'s for which $\ell(q) \geq \ell(p)$. Now suppose that A is $k \times k$ and $k < \ell(p)$. Then every $\mathcal{U}_q(A)$ in (10) vanishes. Hence

(20) $\qquad \mathcal{Y}_p(A) = 0 \quad \text{if } A \text{ is } k \times k \text{ and } k < \ell(p).$

Now write

(21) $\qquad \mathcal{U}_p = \sum_{q \leq p} a^{pq} \mathcal{Y}_q.$

Then

(22) $\qquad \begin{aligned} \mathcal{U}_p(x_1, \ldots, x_k) &= \sum_{q \leq p} a^{pq} \mathcal{Y}_q(x_1, \ldots, x_k) \\ &= \sum_{q \leq p, \ell(q) \leq k} a^{pq} \mathcal{Y}_q(x_1, \ldots, x_k). \end{aligned}$

Similarly

(23) $\qquad \mathcal{Y}_p(x_1, \ldots, x_k) = \sum_{q \leq p, \ell(q) \leq k} a_{pq} \mathcal{U}_q(x_1, \ldots, x_k).$

In view of Lemma 2, (22) and (23) imply that $\{\mathcal{Y}_p, p \in \mathcal{P}_n, \ell(p) \leq k\}$ forms a basis of $V_{n,k}$. ∎

In the proofs of Theorem 3.2.1 and Lemma 3.4.1 we replace all the summations by

(24) $\qquad \sum_{p \in \mathcal{P}_n, \ell(p) \leq k} \quad , \quad \sum_{\substack{q, q' \in \mathcal{P}_n \\ \ell(q) \leq k, \ell(q') \leq k}} \quad etc.$

Then those proofs are complete. We do not repeat the steps of those proofs. But in later proofs we will be careful.

Remark 2. Using the Gale-Ryser theorem (Remark 1), (3.3.2), and (3.3.8) it can be shown that a'_{pq} in (10) is positive if and only if $q \prec p$. This is stronger than Theorem 1.

For future references we record (20) as a corollary.

Corollary 1. *If A is $k \times k$ and $\ell(p) > k$ then $\mathcal{Y}_p(A) = 0$.*

With Theorem 1 we can strengthen the converse part of Theorem 3.1.1.

Theorem 2. *Let integers n, k and a partition $p \in P_n$ be given with $\ell(p) \leq k$. Suppose that f satisfies the following conditions:*

  (i) $f \in V_{n,k}$.

  (ii)   *The leading term in f is the partition p, i.e.*

$$f = \sum_{q \leq p, \ell(q) \leq k} a_{pq}\,_1\mathcal{Y}_q,$$

for some real numbers a_{pq} with $a_{pp} \neq 0$.

  (iii)   *For some constants c_ν,*

$$\mathcal{E}_W f(AW) = c_\nu f(A),$$

for all $k \times k$ symmetric A and for all sufficiently large degrees of freedom ν.

Then $f = a_{pp}\,_1\mathcal{Y}_p$ and $c_\nu = \lambda_{\nu p}$.

Proof. From (ii)

(25) $$\mathcal{E}_W f(AW) = \sum_{q \leq p, \ell(q) \leq k} a_{pq} \lambda_{\nu q}\,_1\mathcal{Y}_q(A).$$

On the other hand by (iii)

(26) $$\mathcal{E}_W f(AW) = c_\nu \sum_{q \leq p, \ell(q) \leq k} a_{pq}\,_1\mathcal{Y}_q(A).$$

Hence

(27) $$0 = \sum_{q \leq p, \ell(q) \leq k} a_{pq}(c_\nu - \lambda_{\nu q})\,_1\mathcal{Y}_q(A)$$

for all $k \times k$ symmetric matrix A. Therefore by Theorem 1 we have $a_{pq}(c_\nu - \lambda_{\nu q}) = 0$ for all q. Considering $q = p$ we obtain $c_\nu = \lambda_{\nu p}$. Then for all $q < p$ we have $a_{pq}(\lambda_{\nu p} - \lambda_{\nu q}) = 0$. Taking $\nu = \nu_0$ we have $a_{pq} = 0$ for all $q < p$. Therefore $f = a_{pp\,1}\mathcal{Y}_p$. This completes the proof. ∎

§ 4.2 EVALUATION OF $_1\mathcal{Y}_p(I_k)$

In the sequel we often work with a normalization denoted by $_1\mathcal{Y}_p$ which has the leading coefficient 1, namely

$$(1) \qquad {}_1\mathcal{Y}_p = \mathcal{U}_p + \sum_{q<p} {}_1 a_{pq} \mathcal{U}_q.$$

Advantages of this normalization will become clear soon.

Remark 1. In several places we already have used the expressions "leading term" or "leading coefficient". Here and in the sequel "leading" refers to the highest partition when a homogeneous symmetric polynomial is expressed in terms of bases $\{\mathcal{U}_p\}$, $\{\mathcal{M}_p\}$, or $\{\mathcal{Y}_p\}$.

We shall evaluate $_1\mathcal{Y}_p(I_k)$. From Theorem 3.2.4 we know that $_1b_p \equiv \lambda_{kp}/{}_1\mathcal{Y}_p(I_k)$ is a constant independent of k. Therefore our goal is to obtain $_1b_p$. Now

$$(2) \qquad {}_1b_p\, {}_1\mathcal{Y}_p(I_k) = \lambda_{kp} = Z_p(I_k).$$

Therefore $_1b_p$ is the leading coefficient of Z_p. This was needed for the unique decomposition of the left hand side of (3.4.11). We use the following recursive relation.

Theorem 1. *If A is a $k \times k$ symmetric matrix, then*

$$(3) \qquad |A|\, {}_1\mathcal{Y}_p(A) = {}_1\mathcal{Y}_{p+(1^k)}(A),$$

where $p + (1^k) = (p_1 + 1, p_2 + 1, \ldots, p_k + 1, p_{k+1}, \ldots) \in \mathcal{P}_{n+k}$, $n = |p|$.

Proof. If $\ell(p) > k$ then ${}_1\mathcal{Y}_p(A) = 0$ by Corollary 4.1.1. In this case $\ell(p + (1^k)) = \ell(p) > k$. Hence ${}_1\mathcal{Y}_{p+(1^k)}(A) = 0$. (3) holds trivially in this case. Now let $\ell(p) \le k$. We can use Theorem 4.1.2. Let $f(A)$ denote the left hand side of (3). Clearly $f \in V_{n+k,k}$ and (i) of Theorem 4.1.2 is verified. With respect to the basis $\{\mathcal{U}_p\}$ the leading term in ${}_1\mathcal{Y}_p$ is $\mathcal{U}_p$. Since $|A|\mathcal{U}_p(A) = \mathcal{U}_{p+(1^k)}(A)$ the leading term in f is $\mathcal{U}_{p+(1^k)}$ and this implies (ii). Now consider

$$(4) \qquad \mathcal{E}_W f(AW) = \mathcal{E}_W\{|AW|\,{}_1\mathcal{Y}_p(AW)\} = |A|\mathcal{E}_W\{|W|\,{}_1\mathcal{Y}_p(AW)\}.$$

Note that we can absorb $|W|$ into the Wishart density which is proportional to $|W|^{(\nu-p-1)/2} \exp(-\frac{1}{2}\operatorname{tr} W)$. This changes the degrees of freedom of the Wishart density, but in any case we have $\mathcal{E}_W\{|W|\,{}_1\mathcal{Y}_p(AW)\} = c_\nu\,{}_1\mathcal{Y}_p(A)$ for some c_ν. (Explicit evaluation of c_ν is straightforward, but we do not need it.) Hence

$$(5) \qquad \mathcal{E}_W f(AW) = |A|\,c_\nu\,{}_1\mathcal{Y}_p(A) = c_\nu f(A)$$

and (iii) is verified. Therefore by Theorem 4.1.2 $|A|\,{}_1\mathcal{Y}_p(A) = c\,{}_1\mathcal{Y}_{p+(1^k)}(A)$ for some c. Comparing the leading term with respect to the basis $\{\mathcal{U}_q\}$ we obtain $c=1$. This completes the proof. ∎

Corollary 1. *(Formula(129) in James (1964))* Let $p = (p_1, \ldots, p_\ell)$ and $p - (p_\ell^\ell) = (p_1 - p_\ell, p_2 - p_\ell, \ldots, p_{\ell-1} - p_\ell)$. Then for an $\ell \times \ell$ symmetric A

$$(6) \qquad {}_1\mathcal{Y}_p(A) = |A|^{p_\ell}\,{}_1\mathcal{Y}_{p-(p_\ell^\ell)}(A).$$

Proof. $|A|^{p_\ell}\,{}_1\mathcal{Y}_{p-(p_\ell^\ell)}(A) = |A|^{p_\ell-1}\,{}_1\mathcal{Y}_{p-(p_\ell^\ell)+(1^\ell)}(A) = \cdots = {}_1\mathcal{Y}_p(A)$. ∎

Applying Corollary 1 to the identity matrices of appropriate dimensionalities we can evaluate ${}_1b_p$ in (2).

Theorem 2.

$$(7) \qquad {}_1b_p = 2^{|p|} \prod_{i=1}^{\ell(p)} \prod_{j=1}^{i} (\frac{1}{2}i - \frac{1}{2}(j-1) + p_j - p_i)_{p_i - p_{i+1}},$$

§ 4.2   Evaluation of $_1\mathcal{Y}_p(I_k)$ 51

where $(a)_k = a(a+1)\cdots(a+k-1)$.

Proof. We prove this by induction on the length of p. Let $\ell(p) = 1$, namely $p = (p_1)$. Then

(8) $$_1\mathcal{Y}_p(I_1) = \mathcal{U}_p(I_1) = 1^{p_1} = 1.$$

Therefore

(9) $$\begin{aligned}_1b_p &= \lambda_{1p}/\,_1\mathcal{Y}_p(I_1) \\ &= 1\cdot 3\cdots(2p_1 - 1) \\ &= 2^{p_1}\left(\frac{1}{2}\right)_{p_1},\end{aligned}$$

which is of the form (7). Now suppose that (7) is true for $\ell(p) = k - 1$. We want to show that then (7) holds for $\ell(p) = k$. Let $p=(p_1,\ldots,p_k)$ and $p - (p_k^k)=(p_1 - p_k, p_2 - p_k, \ldots, p_{k-1} - p_k)$. Note that $\ell(p - (p_k^k)) = k - 1$. Putting I_k in (6) we obtain

(10) $$_1\mathcal{Y}_p(I_k) = \,_1\mathcal{Y}_{p-(p_k^k)}(I_k)$$

or

(11) $$_1b_p = \,_1b_{p-(p_k^k)}\frac{\lambda_{kp}}{\lambda_{k,p-(p_k^k)}}.$$

Using the induction hypothesis
(12)
$$\begin{aligned}
{}_1b_p &= \prod_{i=1}^{k-1}\prod_{j=1}^{i}(\tfrac{1}{2}i-\tfrac{1}{2}(j-1)+(p_j-p_k)-(p_i-p_k))_{(p_i-p_k)-(p_{i+1}-p_k)} \\
&\quad \cdot 2^{|p-(p_k^k)|}\cdot \frac{\lambda_{kp}}{\lambda_{k,p-(p_k^k)}} \\
&= \prod_{i=1}^{k-1}\prod_{j=1}^{i}(\tfrac{1}{2}i-\tfrac{1}{2}(j-1)+p_j-p_i)_{p_i-p_{i+1}} \\
&\quad \cdot 2^{|p-(p_k^k)|}\cdot \frac{2^{|p|}\prod_{j=1}^{k}\Gamma[p_j+\tfrac{1}{2}(k+1-j)]/\Gamma[\tfrac{1}{2}(k+1-j)]}{2^{|p-(p_k^k)|}\prod_{j=1}^{k-1}\Gamma[p_j-p_k+\tfrac{1}{2}(k+1-j)]/\Gamma[\tfrac{1}{2}(k+1-j)]} \\
&= 2^{|p|}\prod_{i=1}^{k-1}\prod_{j=1}^{i}(\tfrac{1}{2}i-\tfrac{1}{2}(j-1)+p_j-p_i)_{p_i-p_{i+1}} \\
&\quad \cdot \prod_{j=1}^{k}((\tfrac{1}{2}k-\tfrac{1}{2}(j-1)+p_j-p_k)_{p_k} \\
&= 2^{|p|}\prod_{i=1}^{k}\prod_{j=1}^{i}(\tfrac{1}{2}i-\tfrac{1}{2}(j-1)+p_j-p_i)_{p_i-p_{i+1}}.
\end{aligned}$$

Therefore (7) holds for $k=\ell(p)$ and the theorem is proved. ∎

There is a curious fact about ${}_1b_p$. Let $k!!$ denote $1\cdot 3\cdots k$ or $2\cdot 4\cdots k$ depending on whether k is odd or even. Then as above it can be shown by induction that

(13) $$\quad {}_1b_p = \prod_{i<j}\frac{(2p_i-2p_j-i+j-2)!!}{(2p_i-2p_j-i+j-1)!!}\prod_{i=1}^{\ell(p)}(2p_i-i+\ell(p)-1)!!.$$

Now $(\operatorname{tr} A)^n = \sum d_p Z_p(A) = \sum d_p\, {}_1b_p\, {}_1Y_p(A)$. From (3.4.12)

(14)
$$\begin{aligned}
d_p\, {}_1b_p &= 2^n n!\prod_{i<j}(2p_i-2p_j-i+j)\Big/\prod_{i=1}^{\ell(p)}(2p_i-i+\ell(p))! \\
&\quad \cdot \prod_{i<j}\frac{(2p_i-2p_j-i+j-2)!!}{(2p_i-2p_j-i+j-1)!!}\prod_{i=1}^{\ell(p)}(2p_i-i+\ell(p)-1)!! \\
&= 2^n n!\prod_{i<j}\frac{(2p_i-2p_j-i+j)!!}{(2p_i-2p_j-i+j-1)!!}\Big/\prod_{i=1}^{\ell(p)}(2p_i-i+\ell(p))!!.
\end{aligned}$$

This is very similar to $_1b_p^{-1}$ if we ignore the constant $2^n n!$. In Section 5.3 we will see that in the complex case the corresponding quantities $\tilde{d}_p$, $_1\tilde{b}_p$ satisfy an exact relation $\tilde{d}_p(_1\tilde{b}_p)^2 = n!$.

§ 4.3 MORE ON INTEGRAL IDENTITIES

In this section we evaluate the constant c_p in Theorem 3.2.2 for several distributions. The first one is the inverted Wishart distribution. See Khatri (1966), Constantine (1963).

Lemma 1. *Let W be distributed according to $\mathcal{W}(I_k, \nu)$, $\nu > 2h(p) + k - 1$. Then for symmetric A*

(1) $$\mathcal{E}_W \mathcal{Y}_p(AW^{-1}) = c_p \mathcal{Y}_p(A),$$

where

(2) $$c_p = \prod_{i=1}^{\ell(p)} \frac{\Gamma[\tfrac{1}{2}(\nu - k + i) - p_i]}{\Gamma[\tfrac{1}{2}(\nu - k + i)] 2^{p_i}}.$$

Proof. Let $A = \text{diag}(\alpha_1, \ldots, \alpha_k)$ without loss of generality. We look at the monomial term $\alpha^{p_1} \cdots \alpha^{p_\ell}$ ($\ell = \ell(p)$). Then as in (3.1.12) its coefficient in (1) is

(3) $$\mathcal{E}_W\{W^{-1}(1)^{p_1-p_2} W^{-1}(1,2)^{p_2-p_3} \cdots W^{-1}(1,\ldots,\ell)^{p_\ell}\},$$

which has to be equal to c_p. Let $W = T'T$ where T is lower triangular with positive diagonal elements. Then analogous to Lemma 3.1.3 t_{ii}, $i = 1, \ldots, k$, are independently distributed according to $\chi(\nu - k + i)$. Then $W^{-1} = T^{-1}T'^{-1}$ and T^{-1} is lower triangular with diagonal elements reciprocal to the diagonal elements of T. Therefore $W^{-1}(1,\ldots,r) = (t_{11} \cdots t_{rr})^{-2}$. Hence

$$c_p = \mathcal{E}\{t_{11}^{-2p_1} \cdots t_{\ell\ell}^{-2p_\ell}\}$$
$$= \prod_{i=1}^{\ell} \frac{\Gamma[\tfrac{1}{2}(\nu - k + i) - p_i]}{\Gamma[\tfrac{1}{2}(\nu - k + i)] 2^{p_i}}$$
$$= \{\prod_{i=1}^{\ell} (\nu - k + i - 2p_i)(\nu - k + i - 2p_i + 2) \cdots (\nu - k + i - 2)\}^{-1}.$$

∎

Related to Lemma 1 we have the following interesting identity which is briefly mentioned in Constantine (1966). Let $p_{s,t}^*$ be defined by (2.1.5).

Lemma 2. *Let A be a $t \times t$ positive definite matrix. Then*

$$|A|^s \frac{\mathcal{Y}_p(A^{-1})}{\mathcal{Y}_p(I_t)} = \frac{\mathcal{Y}_{p_{s,t}^*}(A)}{\mathcal{Y}_{p_{s,t}^*}(I_t)}, \tag{4}$$

where $s \geq h(p)$, $t \geq \ell(p)$.

Proof. Without loss of generality let $A = \text{diag}(\alpha_1, \ldots, \alpha_t)$. let $f(A) = |A|^s \mathcal{Y}_p(A^{-1})$. We use Theorem 4.1.2. In terms of the basis $\{M_q\}$ we can write

$$\mathcal{Y}_p(A^{-1}) = \sum_{q \prec p, \ell(q) \leq t} a_{pq} M_q(1/\alpha_1, \ldots, 1/\alpha_t). \tag{5}$$

Note that $q \prec p$ implies $h(q) \leq h(p)$. Now the degree of $1/\alpha_i$ in $M_q(1/\alpha_1, \ldots, 1/\alpha_t)$ is $h(q)$. Hence the degree of $1/\alpha_i$ in $\mathcal{Y}_p(A^{-1})$ is $h(p)$. Now $|A|^s = (\alpha_1 \cdots \alpha_t)^s$ and $s \geq h(p)$. We see that $1/\alpha_i$ is canceled by $|A|^s$ and $f(A) = |A|^s \mathcal{Y}_p(A^{-1})$ is a polynomial in $(\alpha_1, \ldots, \alpha_t)$. Clearly it is symmetric and homogeneous of degree $st - |p|$. Therefore $f \in V_{st-|p|,t}$. This verifies (i) of Theorem 4.1.2. Now

$$|A|^s M_q(A^{-1}) = (\alpha_1 \cdots \alpha_t)^s \sum_{(i_1,\ldots,i_\ell) \subset (1,\ldots,t)} \frac{1}{\alpha_{i_1}^{q_1} \cdots \alpha_{i_\ell}^{q_\ell}}$$
$$= \sum_{(i_1,\ldots,i_\ell) \subset (1,\ldots,t)} \alpha_{i_1}^{s-q_1} \alpha_{i_2}^{s-q_2} \cdots \alpha_{i_\ell}^{s-q_\ell} \cdot \alpha_{j_1}^s \cdots \alpha_{j_{t-\ell}}^s$$
$$= M_{q_{s,t}^*}(A),$$

where $1 \leq j_1, \ldots, j_{t-\ell} \leq t$ are indices not included in $(i_1, \ldots, i_\ell)$ and $q_{s,t}^* = (s, \ldots, s, s - q_\ell, \ldots, s - q_2, s - q_1)$. Hence by Lemma 2.1.4 the leading term in f is $a_{pp} M_{p_{s,t}^*}$. This verifies (ii). Now consider

$$\mathcal{E}_W f(AW) = \mathcal{E}_W\{|AW|^s \mathcal{Y}_p(A^{-1}W^{-1})\} = |A|^s \mathcal{E}_W\{|W|^s \mathcal{Y}_p(A^{-1}W^{-1})\}.$$

As in the proof of Theorem 4.2.1 $|W|^s$ can be absorbed into the Wishart density and we have $\mathcal{E}_W\{|W|^s \mathcal{Y}_p(A^{-1}W^{-1})\} = c_\nu \mathcal{Y}_p(A^{-1})$ for some c_ν. Therefore

$$\mathcal{E}_W f(AW) = |A|^s c_\nu \mathcal{Y}_p(A^{-1}) = c_\nu f(A).$$

This verifies (iii) and by Theorem 4.1.2 we have $f(A) = |A|^s \mathcal{Y}_p(A^{-1}) = c \mathcal{Y}_{p_{s,t}^*}(A)$ for some c. Putting $A = I_t$ we obtain $c = \mathcal{Y}_p(I_t)/\mathcal{Y}_{p_{s,t}^*}(I_t)$. ∎

The second distribution is a "multivariate F" distribution. There are many ways to generalize the univariate F distribution to the multivariate case. Here we work with the following version. For other generalizations see Johnson and Kotz (1972).

Lemma 3. *Let the columns of $X_1 : k \times \nu_1$, $X_2 : k \times \nu_2$ ($\nu_2 > 2h(p) + k - 1$) be independently distributed according to $\mathcal{N}(0, \Sigma)$. Let $W = X_1'(X_2 X_2')^{-1} X_1$. Then*

$$\mathcal{E}_W \{y_p(AW)\} = \lambda_{kp} \prod_{i=1}^{\ell(p)} \frac{\Gamma[\tfrac{1}{2}(\nu_2 - k + i) - p_i]}{\Gamma[\tfrac{1}{2}(\nu_2 - k + i)] 2^{p_i}} y_p(A). \tag{6}$$

Proof. Premultiplying X_1, X_2 by $\Sigma^{-\tfrac{1}{2}}$ we can take $\Sigma = I_k$ without loss of generality. Then

$$\begin{aligned}
\mathcal{E}_W y_p(AW) &= \mathcal{E}_{X_1} \mathcal{E}_{X_2} y_p(X_1 A X_1' (X_2 X_2')^{-1}) \\
&= \prod_{i=1}^{\ell(p)} \frac{\Gamma[\tfrac{1}{2}(\nu_2 - k + i) - p_i]}{\Gamma[\tfrac{1}{2}(\nu_2 - k + i)] 2^{p_i}} \mathcal{E}_{X_1} y_p(A X_1' X_1) \\
&= \prod_{i=1}^{\ell(p)} \frac{\Gamma[\tfrac{1}{2}(\nu_2 - k + i) - p_i]}{\Gamma[\tfrac{1}{2}(\nu_2 - k + i)] 2^{p_i}} \lambda_{kp} y_p(A).
\end{aligned} \tag{7}$$

∎

Remark 1. It is more or less obvious to prove Lemma 3 for other definitions of multivariate F distribution.

Our last distribution is multivariate beta distribution (Constantine (1963)). The following derivation is essentially the same as in Constantine (1963), but more probabilistic. Let W_1, W_2 be independently distributed according to $\mathcal{W}(\Sigma, \nu_1), \mathcal{W}(\Sigma, \nu_2)$ ($\Sigma : k \times k$) respectively. Note that $W = W_1 + W_2 \sim \mathcal{W}(\Sigma, \nu_1 + \nu_2)$. Now the conditional density of W_1 given W is

$$\begin{aligned}
&f(W_1 \mid W) \\
&= c \frac{|W_1|^{\tfrac{\nu_1 - k - 1}{2}} \exp(-\tfrac{1}{2} \operatorname{tr} \Sigma^{-1} W_1) |W_2|^{\tfrac{\nu_2 - k - 1}{2}} \exp(-\tfrac{1}{2} \operatorname{tr} \Sigma^{-1} W_2)}{|W|^{\tfrac{\nu_1 + \nu_2 - k - 1}{2}} \exp(-\tfrac{1}{2} \operatorname{tr} \Sigma^{-1} W)} \\
&= c \frac{|W_1|^{\tfrac{\nu_1 - k - 1}{2}} |W_2|^{\tfrac{\nu_2 - k - 1}{2}}}{|W|^{\tfrac{\nu_1 + \nu_2 - k - 1}{2}}},
\end{aligned} \tag{8}$$

where

(9) $$c = \frac{\prod_{i=1}^{k} \Gamma[\frac{1}{2}(\nu_1 + \nu_2 - i + 1)]}{\pi^{k(k-1)/4} \prod_{i=1}^{k} \Gamma[\frac{1}{2}(\nu_1 - i + 1)]\Gamma[\frac{1}{2}(\nu_2 - i + 1)]}.$$

Note that terms involving Σ cancel out in (8). Therefore the conditional distribution does not depend on Σ. When $W=I$, $f(W_1 \mid I)$ is called multivariate beta density:

(10) $$f(W_1 \mid I) = c|W_1|^{\frac{\nu_1-k-1}{2}}|I - W_1|^{\frac{\nu_2-k-1}{2}}.$$

Since this density is orthogonally invariant the conditional distribution of W_1 given $W = I$ is orthogonally invariant. Now we want to evaluate c_p in

(11) $$\mathcal{E}\{Y_p(AW_1) \mid W = I\} = c_p Y_p(A).$$

For a positive definite A let $A^{\frac{1}{2}} = \Gamma D^{\frac{1}{2}} \Gamma'$ where Γ is orthogonal and D is diagonal in $A = \Gamma D \Gamma'$. Now the conditional distribution of $A^{\frac{1}{2}} W_1 A^{\frac{1}{2}}$ given $W = I$ is the same as the conditional distribution of W_1 given $W = A$. This follows from the above mentioned fact that the conditional distribution does not depend on Σ. Therefore

(12) $$\mathcal{E}\{Y_p(AW_1) \mid W = I\} = \mathcal{E}\{Y_p(W_1) \mid W = A\}.$$

Letting $A = W_1 + W_2$ we obtain from (11) and (12)

(13) $$\mathcal{E}\{Y_p(W_1) \mid W_1 + W_2\} = c_p Y_p(W_1 + W_2).$$

Now taking unconditional expectation we obtain

(14) $$\lambda_{\nu_1 p} Y_p(\Sigma) = c_p \lambda_{\nu_1+\nu_2,p} Y_p(\Sigma).$$

Hence $c_p = \lambda_{\nu_1 p}/\lambda_{\nu_1+\nu_2,p}$. Now we have proved

Lemma 4. *Let W_1 have the density (10). Then*

(15) $$\mathcal{E}_{W_1} \mathcal{Y}_p(AW_1) = \frac{\lambda_{\nu_1 p}}{\lambda_{\nu_1+\nu_2,p}} \mathcal{Y}_p(A).$$

Variations of the above three lemmas can be found in Khatri (1966), Subrahmaniam (1976).

§ 4.4 COEFFICIENTS OF $\mathcal{U}_q$ IN $\mathcal{Y}_p$

In this section we study coefficients of $\mathcal{U}_q$'s when zonal polynomials are expressed as linear combinations of $\mathcal{U}_q$'s. For definiteness we work with ${}_1a_{pq}$ in ${}_1\mathcal{Y}_p = \mathcal{U}_p + \sum {}_1a_{pq} \mathcal{U}_q$. If rank $A=1,2$ all the relevant coefficients are known and we can compute $\mathcal{Y}_p(A)$ explicitly. We review this first. After that we study several recurrence relations between the coefficients. When rank $A > 2$ these recurrence relations are not enough to compute the values of zonal polynomials $\mathcal{Y}_p(A)$ for all p. Nonetheless they seem to be very useful. Coefficients of $\mathcal{M}_q$'s will be discussed in the next section and $\mathcal{T}_q$'s in Section 4.6. We discuss relative advantages of various bases on the way.

4.4.1 Rank 1 and rank 2 cases

If A is symmetric and rank $A=1$ then A has only one nonzero root. Let $A = \text{diag}(\alpha_1, 0, \ldots, 0)$ without loss of generality. By Corollary 4.1.1 $\mathcal{Y}_p(A) = 0$ if $\ell(p) \geq 2$. Therefore only onepart partitions $p = (p_1)$ count. Obviously

(1) $${}_1\mathcal{Y}_{(p_1)}(A) = \mathcal{U}_{(p_1)}(A) = \alpha_1^{p_1}.$$

Therefore in this case zonal polynomials reduce to powers of α_1.

Now suppose rank $A=2$. Let $A=\text{diag}(\alpha_1, \alpha_2, 0, \ldots, 0)$ and $\tilde{A}=\text{diag}(\alpha_1, \alpha_2)$: 2×2. We have to consider only partitions with two parts $p = (p_1, p_2)$. Now we use Corollary 4.2.1:

(2) $${}_1\mathcal{Y}_{(p_1,p_2)}(A) = {}_1\mathcal{Y}_{(p_1,p_2)}(\tilde{A}) = |\tilde{A}|^{p_2} {}_1\mathcal{Y}_{(p_1-p_2)}(\tilde{A}),$$

where (p_1-p_2) is a onepart partition. Therefore it suffices to know the value of a zonal polynomial of onepart partition evaluated at 2×2 matrix $\tilde{A}$. Actually zonal polynomials of onepart partitions are known explicitly and can be derived as follows. If we let $\beta_1 = 1, \beta_2 = \cdots = \beta_k = 0$ in (3.4.6) we obtain

$$(3) \qquad \prod_{i=1}^{k}(1-2\theta\alpha_i)^{-\frac{1}{2}} = \sum_{n=0}^{\infty}(\theta^n/n!)d_{(n)}Z_{(n)}(A)\,_1b_{(n)},$$

where $_1b_{(n)} = Z_{(n)}(I_1)$ is the leading coefficient of $Z_{(n)}$ (see (4.2.2)). Note that $(\operatorname{tr} C)^n = \mathcal{U}_{(n)}(C)$ and with respect to the basis $\{\mathcal{U}_p\}$ only $Z_{(n)}(C)$ contains $\mathcal{U}_{(n)}$. Therefore in (3.4.1) $(\operatorname{tr} C)^n = {_1}\mathcal{Y}_{(n)}(C) + \cdots = (1/{_1}b_{(n)}) Z_{(n)}(C) + \cdots$. Hence $d_{(n)} = {_1}b_{(n)}^{-1}$. We have

$$(4) \qquad \prod_{i=1}^{k}(1-2\theta\alpha_i)^{-\frac{1}{2}} = \sum_{n=0}^{\infty}(\theta^n/n!)Z_{(n)}(A).$$

The left hand side can be expanded as follows.

$$(5) \quad \begin{aligned}\prod_{i=1}^{k}(1-2\theta\alpha_i)^{-\frac{1}{2}} &= \{1-(2\theta u_1 - 4\theta^2 u_2 + \cdots)\}^{-\frac{1}{2}} \\ &= \sum_{n=0}^{\infty}(2\theta)^n \sum_{p \in P_n} \frac{1}{p_1!}\frac{1}{2}\frac{3}{2}\cdots\frac{2p_1-1}{2}\binom{p_1}{p_1-p_2, p_2-p_3, \ldots, p_{\ell(p)}} \\ &\quad \cdot (-1)^{(p_2-p_3)+(p_4-p_5)+\cdots}\mathcal{U}_p(A).\end{aligned}$$

This follows from the fact that $\mathcal{U}_p$ being a product of p_1 terms comes only from p_1-th power term in the expansion of $(1-2\theta u_1 + \cdots)^{-1/2}$. Comparing (4) and (5) we obtain

$$(6) \qquad Z_{(n)} = 2^n n! \sum_{p \in P_n}(-1)^{(p_2-p_3)+(p_4-p_5)+\cdots}\frac{(\frac{1}{2})_{p_1}}{(p_1-p_2)!\cdots p_{\ell(p)}!}\mathcal{U}_p.$$

Note that $_1b_{(n)} = 1 \cdot 3 \cdots (2n-1)$, $2^n n! = 2 \cdot 4 \cdots (2n)$, $|\tilde{A}|^{p_2}\mathcal{U}_{(q_1,q_2)}(\tilde{A}) = \mathcal{U}_{(q_1+p_2, q_2+p_2)}(\tilde{A}) = \mathcal{U}_{(q_1+p_2, q_2+p_2)}(A)$. Therefore combining these equalities

§ 4.4 *Coefficients of U_q*

we obtain

(7)
$$_1\mathcal{Y}_{(p_1,p_2)}(A) = \frac{2\cdot 4\cdots(2p_1-2p_2)}{1\cdot 3\cdots(2p_1-2p_2-1)}$$
$$\cdot \sum_{(q_1,q_2)\in P_{p_1-p_2}} (-1)^{q_2} \frac{(\frac{1}{2})_{q_1}}{(q_1-q_2)! q_2!} U_{(q_1+p_2,q_2+p_2)}(A).$$

if rank $A = 2$. See formula (130) in James (1964).

If rank $A=3$ what we have to know are the values of zonal polynomials of twopart partitions evaluated at a rank 3 matrix A. Obviously things become more and more complicated as rank A increases. However several useful recurrence relations on the coefficients can be obtained.

4.4.2 Recurrence relations on the coefficients

We present here three recurrence relations. The first one has been already used in deriving (7).

Lemma 1. *If $k \geq \ell(p), k \geq \ell(q)$, then*

(8)
$$_1a_{pq} = {}_1a_{p+(1^k),q+(1^k)}.$$

Proof. Let A be $k \times k$. Then

(9)
$$|A|\,_1\mathcal{Y}_p(A) = |A|\{U_p(A) + \sum_{q<p, \ell(q)\leq k} {}_1a_{pq} U_q(A)\}$$
$$= U_{p+(1^k)}(A) + \sum_{q<p,\ell(q)\leq k} {}_1a_{pq} U_{q+(1^k)}(A).$$

By Theorem 4.2.1

(10)
$$|A|\,_1\mathcal{Y}_p(A) = {}_1\mathcal{Y}_{p+(1^k)}(A)$$
$$= U_{p+(1^k)}(A) + \sum_{q'<p+(1^k),\ell(q')\leq k} {}_1a_{p+(1^k),q'} U_{q'}(A).$$

Comparing (9) and (10) we obtain by Lemma 4.1.2 $\quad _1a_{pq} = {}_1a_{p+(1^k),q+(1^k)}.$

∎

Remark 1. Theorem 4.2.1 has been known and Lemma 1 is almost an immediate consequence. However it does not seem to have been explicitly stated.

The next one is in a sense conjugate to Lemma 1. Let $p = (p_1, \ldots, p_\ell) \in P_n$ and $m \geq p_1 = h(p)$. We denote by (m,p) the partition $(m, p_1, p_2, \ldots, p_\ell) \in P_{n+m}$.

Theorem 1. *Let $m \geq h(p)$. Then*

$$(11) \qquad \frac{\partial^m}{\partial \alpha_{k+1}^m} {}_1\mathcal{Y}_{(m,p)}(\alpha_1, \ldots, \alpha_{k+1}) = m! \, {}_1\mathcal{Y}_p(\alpha_1, \ldots, \alpha_k).$$

Proof. With respect to the basis $\{\mathcal{U}_p\}$ let

$$(12) \quad \begin{aligned} {}_1\mathcal{Y}_{(m,p)}&(\alpha_1, \ldots, \alpha_{k+1}) \\ &= \sum_{q \leq (m,p), q \in P_{n+m}} {}_1a_{(m,p),q} \mathcal{U}_q(\alpha_1, \ldots, \alpha_{k+1}) \\ &= \sum_{(m,q') \leq (m,p), q' \in P_n} {}_1a_{(m,p),(m,q')} \mathcal{U}_{(m,q')}(\alpha_1, \ldots, \alpha_{k+1}) \\ &\quad + \sum_{q \in P_{n+m}, h(q) < m} {}_1a_{(m,p),q} \mathcal{U}_q(\alpha_1, \ldots, \alpha_{k+1}). \end{aligned}$$

We differentiate (12) m times with respect to α_{k+1}. Now the degree of α_{k+1} in

$$(13) \qquad \mathcal{U}_q(\alpha_1, \ldots, \alpha_{k+1}) = (\sum \alpha_i)^{q_1-q_2} (\sum_{i<j} \alpha_i \alpha_j)^{q_2-q_3} \cdots$$

is $q_1 = (q_1 - q_2) + (q_2 - q_3) + \cdots + q_{\ell(q)}$. Therefore the terms in the second summation on the right hand side of (12) drop out. Now $\mathcal{U}_{(m,q')}(\alpha_1, \ldots, \alpha_{k+1})$ is a product of $m = (m - q'_1) + \cdots + q'_{\ell(q')}$ elementary symmetric functions $u_r(\alpha_1, \ldots, \alpha_{k+1})$ which are linear in α_{k+1}. Therefore differentiating $\mathcal{U}_{(m,q')}$ m times we are left with the term where each u_r is differentiated exactly once. Furthermore

$$(14) \qquad \frac{\partial}{\partial \alpha_{k+1}} u_r(\alpha_1, \ldots, \alpha_{k+1}) = u_{r-1}(\alpha_1, \ldots, \alpha_k).$$

§ 4.4   *Coefficients of* $\mathcal{U}_q$ 61

Therefore by the chain rule of differentiation

(15)
$$\frac{\partial^m}{\partial \alpha_{k+1}^m} \mathcal{U}_{(m,q')}(\alpha_1,\ldots,\alpha_{k+1})$$
$$= m!\{\frac{\partial}{\partial \alpha_{k+1}} u_1(\alpha_1,\ldots,\alpha_{k+1})\}^{m-q'_1}\{\frac{\partial}{\partial \alpha_{k+1}} u_2(\alpha_1,\ldots,\alpha_{k+1})\}^{q'_1-q'_2}\cdots$$
$$= m! u_1(\alpha_1,\ldots,\alpha_k)^{q'_1-q'_2} u_2(\alpha_1,\ldots,\alpha_k)^{q'_2-q'_3}\cdots$$
$$= m!\mathcal{U}_{q'}(\alpha_1,\ldots,\alpha_k).$$

Let $f(\alpha_1,\ldots,\alpha_k) = (\partial^m/\partial \alpha_{k+1}^m) {}_1\mathcal{Y}_{(m,p)}(\alpha_1,\ldots,\alpha_{k+1})$. Then we have

(16)
$$f(\alpha_1,\ldots,\alpha_k) = \sum_{q\leq p} {}_1 a_{(m,p),(m,q)} \frac{\partial^m}{\partial \alpha_{k+1}^m} \mathcal{U}_{(m,q)}(\alpha_1,\ldots,\alpha_{k+1})$$
$$= m! \sum_{q\leq p} {}_1 a_{(m,p),(m,q)} \mathcal{U}_q(\alpha_1,\ldots,\alpha_k).$$

We replaced q' by q and $(m,q) \leq (m,p)$ by $q \leq p$ since $(m,q) \leq (m,p)$ if and only if $q \leq p$. We have shown that conditions (i) and (ii) of Theorem 4.1.2 are satisfied. We are going to show that condition (iii) of Theorem 4.1.2 is satisfied as well.

Let $\mathbf{A} = \mathrm{diag}(\alpha_1,\ldots,\alpha_{k+1})$ and $\mathbf{A}_1 = \mathrm{diag}(\alpha_1,\ldots,\alpha_k)$. Then exactly as above we obtain

(17)
$$\frac{\partial^m}{\partial \alpha_{k+1}^m} {}_1\mathcal{Y}_{(m,p)}(\mathbf{AW})$$
$$= m! \sum_{q\leq p} {}_1 a_{(m,p),(m,q)} \mathbf{W}(k+1)^{m-q_1} (\sum_{i_1}^{k} \alpha_{i_1} \mathbf{W}(i_1, k+1))^{q_1-q_2}\cdots$$
$$\cdot (\sum_{i_1<\cdots<i_{\ell(q)}}^{k} \alpha_{i_1}\cdots\alpha_{i_{\ell(q)}} \mathbf{W}(i_1,\ldots,i_{\ell(q)}, k+1))^{q_{\ell(q)}}.$$

Let $\mathbf{W}$ be partitioned as

(18)
$$\mathbf{W} = \begin{pmatrix} \mathbf{W}_{11} & \mathbf{w}_{k+1} \\ \mathbf{w}'_{k+1} & w_{k+1,k+1} \end{pmatrix},$$

where $w_{k+1,k+1} = W(k+1)$ is a scalar. Let

(19) $$W_{11 \cdot k+1} = W_{11} - w_{k+1}w'_{k+1}/w_{k+1,k+1}.$$

Then by the well known identity on the determinant of partitioned matrices we have

(20) $$W(i_1, \ldots, i_r, k+1) = w_{k+1,k+1} W_{11 \cdot k+1}(i_1, \ldots, i_r).$$

Therefore in (17) $w_{k+1,k+1}^m = w_{k+1,k+1}^{(m-q_1)+(q_1-q_2)+\cdots}$ comes out as a common factor and we obtain

(21) $$\begin{aligned}\frac{\partial^m}{\partial \alpha_{k+1}^m} {}_1\mathcal{Y}_{(m,p)}(AW) &= m! \sum_{q \leq p} {}_1 a_{(m,p),(m,q)} w_{k+1,k+1}^m \mathcal{U}_q(A_1 W_{11 \cdot k+1}) \\ &= w_{k+1,k+1}^m f(A_1 W_{11 \cdot k+1}).\end{aligned}$$

Now if W is distributed according to $\mathcal{W}(I_{k+1}, \nu)$ then $w_{k+1,k+1}$ and $W_{11 \cdot k+1}$ are independently distributed according to $\chi^2(\nu)$, $\mathcal{W}(I_k, \nu - 1)$ respectively. (See Srivastava and Khatri (1979), Theorem 3.3.5 or Mardia, Kent, and Bibby (1979), Theorem 3.4.6.) Therefore taking expectation with respect to W

(22) $$\begin{aligned}\lambda_{\nu,(m,p)} f(A_1) &= \lambda_{\nu,(m,p)} \frac{\partial^m}{\partial \alpha_{k+1}^m} {}_1 \mathcal{Y}_{(m,p)}(\alpha_1, \ldots, \alpha_{k+1}) \\ &= \mathcal{E}_W\{\frac{\partial^m}{\partial \alpha_{k+1}^m} {}_1 \mathcal{Y}_{(m,p)}(AW)\} \\ &= \mathcal{E}_W\{w_{k+1,k+1}^m f(A_1 W_{11 \cdot k+1})\} \\ &= \mathcal{E}(w_{k+1,k+1}^2) \cdot \mathcal{E}_W\{f(A_1 W_{11 \cdot k+1})\}.\end{aligned}$$

Letting $c_{\nu-1} = \lambda_{\nu,(m,p)}/\mathcal{E}(w_{k+1,k+1}^2)$ we have

(23) $$\mathcal{E} f(A_1 W_{11 \cdot k+1}) = c_{\nu-1} f(A_1)$$

for all $k \times k$ symmetric A_1 and for all sufficiently large ν. This verifies condition (iii) of Theorem 4.1.2 and we conclude

(24) $$f = c_1 \mathcal{Y}_p$$

for some c. Comparing the leading coefficient with respect to the bases $\{\mathcal{U}_p\}$ we obtain $c = m!$. ∎

§ 4.4   Coefficients of $\mathcal{U}_q$

Corollary 1.   *If $m \geq h(p)$ then*

(25) $$_1a_{pq} = {_1a_{(m,p),(m,q)}}.$$

Proof.   From (11) and (15)

(26)
$$\begin{aligned} m!(\mathcal{U}_p + \sum_{q<p} {_1a_{pq}}\mathcal{U}_q) \\ = m!\,_1\mathcal{Y}_p &= \frac{\partial^m}{\partial \alpha_{k+1}^m}\, _1\mathcal{Y}_{(m,p)} \\ &= m!(\mathcal{U}_p + \sum_{q<p} {_1a_{(m,p),(m,q)}}\mathcal{U}_q). \end{aligned}$$

Therefore (25) holds. ∎

In terms of the diagram of p Lemma 1 corresponds to adding a column to the left of the diagram and Corollary 1 corresponds to adding a row to the top. In this sense they are "conjugate". It might be interesting to interpret this result from group representation theory.

```
X  .  .  .  .           X  X  X  X  X
X  .  .                 .  .  .  .
X  .  .                 .  .
X  .                    .  .
X                       .
```

Figure 4.1.

Our third recurrence relation follows from Lemma 4.3.2.

Lemma 2.   *If $s \geq h(p)$, $s \geq h(q)$, $t \geq \ell(p)$, $t \geq \ell(q)$, then*

(27) $$_1a_{pq} = {_1a_{p^*_{s,t}, q^*_{s,t}}}.$$

Proof.   Let A be a $t \times t$ positive definite matrix. From Lemma 4.3.2

(28) $$|A|^s\, _1\mathcal{Y}_p(A^{-1}) = c\, _1\mathcal{Y}_{p^*_{s,t}}(A),$$

or

$$\begin{aligned}(29)\quad &|A|^s U_p(A^{-1}) + \sum_{q<p,\ell(q)\leq t} {}_1a_{pq}|A|^s U_q(A^{-1}) \\ &= c(U_{p^*}(A) + \sum_{q'<p^*,\ell(q')\leq t} {}_1a_{p^*q'}U_{q'}(A)),\end{aligned}$$

where $p^* = p^*_{s,t}$ and $c = {}_1\mathcal{Y}_p(I_t)/{}_1\mathcal{Y}_{p^*}(I_t)$. Now let $A = \mathrm{diag}(\alpha_1, \ldots, \alpha_t)$. Then

$$\begin{aligned}(30)\quad |A|u_r(A^{-1}) &= (\alpha_1 \cdots \alpha_t) \sum_{i_1<\cdots<i_r}^{t} \frac{1}{\alpha_{i_1}\cdots\alpha_{i_r}} \\ &= \sum_{j_1<\cdots<j_{t-r}}^{t} \alpha_{j_1}\cdots\alpha_{j_{t-r}} \\ &= u_{t-r}(A).\end{aligned}$$

Note that (30) is true for $r = 0, t$ if we define $u_0 = 1$. Therefore

$$\begin{aligned}(31)\quad |A|^s U_q(A^{-1}) &= |A|^{s-q_1}\{|A|^{q_1}u_1(A^{-1})^{q_1-q_2}\cdots u_{\ell(q)}(A^{-1})^{q_{\ell(q)}}\} \\ &= u_t(A)^{s-q_1}u_{t-1}(A)^{q_1-q_2}\cdots u_{t-\ell(q)}(A)^{q_{\ell(q)}} \\ &= U_{q^*_{s,t}}(A),\end{aligned}$$

because $q^*_{s,t} = (s,\ldots,s,s-q_{\ell(q)},\ldots,s-q_2,s-q_1)$ and $\ell(q^*_{s,t}) = t$. Substituting (31) into (29) we obtain

$$\begin{aligned}(32)\quad &U_{p^*}(A) + \sum_{q<p,\ell(q)\leq t} {}_1a_{pq}U_{q^*_{s,t}}(A) \\ &= c\{U_{p^*}(A) + \sum_{q'<p^*,\ell(q')\leq t} {}_1a_{p^*q'}U_{q'}(A)\}.\end{aligned}$$

Therefore by Lemma 4.1.2 ${}_1a_{pq} = {}_1a_{p^*_{s,t},q^*_{s,t}}$, $c=1$. ∎

Remark 2. Again this lemma is much easier to grasp in terms of the diagram. See Figure 2.2.

§ 4.5   *Coefficients of* M_q 65

Looking at Table 2 in Parkhurst and James (1974) we find that the above three recurrence relations give a large number of coefficients without any calculation (except that the table is for Z_p rather than for $_1\mathcal{Y}_p$). However it is not clear whether this kind of approach can be further carried out to give all coefficients of zonal polynomials.

§ 4.5 COEFFICIENTS OF M_q

So far we have been mainly working with $\mathcal{U}_p$'s. But in view of Lemma 2.2.2, Lemma 4.1.1 etc. we could have worked with M_p's as well. We defined zonal polynomials in connection with the Wishart distribution and it was more straightforward to define zonal polynomials in terms of $\mathcal{U}_p$'s in that setting. But when it comes to obtaining coefficients it seems easier to work with M_p's. In this section we translate every result in Section 4.4 into the coefficients of M_p's. Another big advantage of working with monomial symmetric functions is a partial differential equation by James (1968), from which he derived a recurrence relation on the coefficients of monomial symmetric functions in a zonal polynomial. (Note that the recurrence relations of Section 4.4.2 were on the coefficients of $\mathcal{U}_q$'s in different zonal polynomials. Here the recurrence relation is on the coefficients in one zonal polynomial.) Actually it is possible to develop a whole theory of zonal polynomials from the partial differential equation. This is done in a recent book by Muirhead (1982) explicitly and illustratively following James (1968). We discuss the partial differential equation and the recurrence relation in Section 4.5.4.

Furthermore Jacob Towber (personal communication) has recently developed a combinatorial method for determining the coefficients. His method involves several steps of counting related to the diagram of a partition. At the moment the combinatorics involved seems to be too complicated to obtain an explicit formula for the coefficients, but it might be carried out.

From the above discussion we see that we have much more information on the coefficients of M_p's than on the coefficients of $\mathcal{U}_p$'s. Therefore in a sense it is pointless to work with $\mathcal{U}_p$'s any more. However from a computational point of view it is easier to compute $\mathcal{U}_q$'s once we obtain the characteristic roots and the characteristic equation of a matrix A. We simply multiply the elementary symmetric functions. In the case of M_p's we have to multiply the roots in all possible ways and sum them up. The relative advantages of M_p's and U_p's should be judged from this viewpoint too.

4.5.1 Rank 1 and rank 2 cases

Let $p = (n)$ be a onepart partition. To express $Z_{(n)}$ in monomial symmetric functions we can use the integral representation by Kates. This was done by Kates (1980). Letting $r = 1$ in (3.3.8) we obtain

$$
(1) \qquad UAU'(1) = \sum_{i=1}^{k} \alpha_i u_{1i}^2,
$$

where $A = \mathrm{diag}(\alpha_1, \ldots, \alpha_k)$ and u_{1i}, $i = 1, \ldots, k$, are independent standard normal variables. Therefore by (3.3.2)

$$
(2) \qquad Z_{(n)} = \mathcal{E}(\sum_{i=1}^{k} \alpha_i u_{1i}^2)^n.
$$

Now the coefficient of $\alpha_1^{p_1} \cdots \alpha_\ell^{p_\ell}$ on the right hand side is

$$
(3) \qquad \binom{n}{p_1, p_2, \ldots, p_\ell} \mathcal{E}\{u_{11}^{2p_1} \cdots u_{1\ell}^{2p_\ell}\} = \frac{n!}{p_1! \cdots p_\ell!} \frac{(2p_1)!}{2^{p_1} p_1!} \cdots \frac{(2p_\ell)!}{2^{p_\ell} p_\ell!}
$$
$$
= n! 2^{-n} \binom{2p_1}{p_1} \cdots \binom{2p_\ell}{p_\ell}.
$$

Therefore

$$
(4) \qquad Z_{(n)} = n! 2^{-n} \sum_{p \in \mathcal{P}_n} M_p(A) \prod_{i=1}^{\ell(p)} \binom{2p_i}{p_i}.
$$

This looks nicer than (4.4.6). $_1\mathcal{Y}_p$ has the leading coefficient 1, so

$$
(5) \qquad {}_1\mathcal{Y}_{(n)} = \binom{2n}{n}^{-1} \sum_{p \in \mathcal{P}_n} M_p \prod_{i=1}^{\ell(p)} \binom{2p_i}{p_i}.
$$

Now let $A = \text{diag}(\alpha_1, \alpha_2)$. Then for $q=(q_1,q_2)$ $(q_1 \neq q_2)$

(6)
$$\begin{aligned}|A|^k M_q(A) &= (\alpha_1\alpha_2)^k(\alpha_1^{q_1}\alpha_2^{q_2} + \alpha_2^{q_1}\alpha_1^{q_2}) \\ &= \alpha_1^{q_1+k}\alpha_2^{q_2+k} + \alpha_2^{q_1+k}\alpha_1^{q_2+k} \\ &= M_{(q_1+k, q_2+k)}(A).\end{aligned}$$

(The equality of the extreme left and the extreme right hand sides holds for $q_1=q_2$ too). Therefore from (4.4.2) we obtain

(7)
$$\begin{aligned}&{}_1\mathcal{Y}_{(p_1,p_2)}(\alpha_1,\alpha_2) \\ &= \binom{2p_1-2p_2}{p_1-p_2}^{-1} \sum_{(q_1,q_2)\in P_{p_1-p_2}} \binom{2q_1}{q_1}\binom{2q_2}{q_2} M_{(q_1+p_2, q_2+p_2)}(\alpha_1, \alpha_2).\end{aligned}$$

This takes care of rank 1 and rank 2 cases.

4.5.2 Again on the generating function of zonal polynomials

To express T_p in terms of M_q's we can simply expand T_p and count various monomial terms. Therefore it seems easier to express the right hand side of (3.4.9) in M_q's than in U_q's. Then we decompose the resulting positive definite coefficient matrix as LL' where L is lower triangular with positive diagonal elements. The elements of L give the desired coefficients of zonal polynomials. The development on Section 3.4 goes through in exactly the same way except that we order $\{T_p, p \in P_n\}$ according to the lexicographic ordering of the conjugate partition p' (see Remark 2.2.2). We do not repeat it here.

Rather we notice here the similarity between two generating functions (3.4.6) and (4.4.4). Let $\gamma_1, \ldots, \gamma_{k^2}$ denote the k^2 numbers $\alpha_i\beta_j$, $i=1,\ldots,k$, $j=1,\ldots,k$. Let $C=\text{diag}(\gamma_1, \ldots, \gamma_{k^2})$. Then from (3.4.6) and (4.4.4) we have

(8)
$$\sum_{n=0}^{\infty}(\theta^n/n!)\sum_{p\in P_n} d_p Z_p(A) Z_p(B) = \sum_{n=0}^{\infty}(\theta^n/n!) Z_{(n)}(C).$$

Hence

(9)
$$\sum_{p\in P_n} d_p Z_p(A) Z_p(B) = Z_{(n)}(C).$$

Now we can substitute (4) into the right hand side. Then it reduces to expressing $M_p(C)$ as a sum of products $M_q(A)M_{q'}(B)$. This seems nicer than directly expanding the right hand side of (3.4.9).

Finally we prove that the coefficient of $M_{(1^k)}$ in Z_p, $p \in P_k$ is $k!$. This is stated in James (1968).

Lemma 1. *Let $p \in P_k$ and $A = \mathrm{diag}(\alpha_1, \ldots, \alpha_k)$. Then*

$$(10) \qquad \frac{\partial^k}{\partial \alpha_1 \partial \alpha_2 \cdots \partial \alpha_k} Z_p(A) = k!.$$

Hence the coefficient of $M_{(1^k)}$ in Z_p is $k!$.

Proof.

$$(11) \qquad \prod_{i,j}^{k} (1 - 2\theta \alpha_i \beta_j)^{-\frac{1}{2}} = \sum_{n=0}^{\infty} (\theta^n/n!) \sum_{p \in P_n} d_p Z_p(A) Z_p(B).$$

Differentiating this by $\alpha_1, \alpha_2, \ldots, \alpha_k$ we obtain

$$(12) \qquad \begin{aligned} &\prod_{i=1}^{k} \left(\sum_{j=1}^{k} \frac{\theta \beta_j}{1 - 2\theta \alpha_i \beta_j} \right) \prod_{i,j}^{k} (1 - 2\theta \alpha_i \beta_j)^{-\frac{1}{2}} \\ &= \sum_{n=0}^{\infty} (\theta^n/n!) \sum_{p \in P_n} d_p Z_p(B) \frac{\partial^k}{\partial \alpha_1 \cdots \partial \alpha_k} Z_p(A). \end{aligned}$$

Now $\theta \beta_j / (1 - 2\theta \alpha_i \beta_j) = \theta \beta_j + \textit{higher order in } \theta$. Hence

$$(13) \qquad \prod_{i=1}^{k} \left(\sum_{j=1}^{k} \frac{\theta \beta_j}{1 - 2\theta \alpha_i \beta_j} \right) = \theta^k (\sum_{j=1}^{k} \beta_j)^k + \textit{higher order in } \theta.$$

Comparing the coefficients of θ^k we obtain

$$(14) \qquad (\sum_{j=1}^{k} \beta_j)^k = \sum_{p \in P_k} \frac{d_p}{k!} Z_p(B) \frac{\partial^k}{\partial \alpha_1 \cdots \partial \alpha_k} Z_p(A).$$

§ 4.5   Coefficients of M_q 69

But by (3.4.1)

(15) $$(\sum_{j=1}^{k} \beta_j)^k = (\operatorname{tr} \boldsymbol{B})^k = \sum_{p \in P_k} d_p Z_p(\boldsymbol{B}).$$

Comparing (14) and (15) we obtain

$$\frac{\partial^k}{\partial \alpha_1 \cdots \partial \alpha_k} Z_p(\boldsymbol{A}) = k!$$

∎

4.5.3 Recurrence relations of Section 4.4.2

Here we work again with the normalization $_1\mathcal{Y}_p$. Let

(16) $$_1\mathcal{Y}_p = M_p + \sum_{q < p} {}_1 b_{pq} M_q.$$

Lemma 2.   If $k \geq \ell(p), k \geq \ell(q)$, then

(17) $$_1 b_{pq} = {}_1 b_{p+(1^k), q+(1^k)}.$$

Proof.   Let $\boldsymbol{A} = \operatorname{diag}(\alpha_1, \ldots, \alpha_k)$ where $k \geq \ell(p)$. Then

(18) $$\begin{aligned} |\boldsymbol{A}| M_q(\boldsymbol{A}) &= (\alpha_1 \cdots \alpha_k) \sum_{(i_1,\ldots,i_\ell) \subset (1,\ldots,k)} \alpha_{i_1}^{q_1} \alpha_{i_2}^{q_2} \cdots \alpha_{i_\ell}^{q_\ell} \\ &= \sum_{(i_1,\ldots,i_\ell) \subset (1,\ldots,k)} \alpha_{i_1}^{q_1+1} \alpha_{i_2}^{q_2+1} \cdots \\ &= M_{q+(1^k)}(\boldsymbol{A}). \end{aligned}$$

Note that this equality does <u>not</u> hold for augmented monomial symmetric functions. The equality above holds because the summation is over distinguishable terms and $\alpha_{i_1}^{q_1} \cdots \alpha_{i_\ell}^{q_\ell}$ is distinguishable from $\alpha_{j_1}^{q_1} \cdots \alpha_{j_\ell}^{q_\ell}$ if and only if $(\alpha_1 \cdots \alpha_k) \alpha_{i_1}^{q_1} \cdots \alpha_{i_\ell}^{q_\ell}$ is distinguishable from $(\alpha_1 \cdots \alpha_k) \alpha_{j_1}^{q_1} \cdots \alpha_{j_\ell}^{q_\ell}$. For augmented monomial functions refer to (2.2.6). Now the lemma can be proved just as Lemma 4.4.1 if we replace $\mathcal{U}_p, \mathcal{U}_q, {}_1 a_{pq}$ in (4.4.9) by $M_p, M_q, {}_1 b_{pq}$ respectively.

∎

Lemma 3. *Let $h(p) \leq m$. Then*

(19) $$\mathstrut_1 b_{pq} = \mathstrut_1 b_{(m,p),(m,q)}.$$

Proof. The degree of α_{k+1} in $M_q(\alpha_1, \ldots, \alpha_{k+1})$ is $h(q)$. Hence if $h(q) < m$, then

(20) $$\frac{\partial^m}{\partial \alpha_{k+1}^m} M_q(\alpha_1, \ldots, \alpha_{k+1}) = 0.$$

If $h(q) = m$ let $q = (m, q')$. Then clearly

(21) $$\frac{\partial^m}{\partial \alpha_{k+1}^m} M_q(\alpha_1, \ldots, \alpha_{k+1}) = m! M_{q'}(\alpha_1, \ldots, \alpha_k).$$

(Again this equality does not hold for AM_q.) Now (4.4.26) holds with M_p, M_q, $\mathstrut_1 b_{pq}$ replacing $U_p, U_q, \mathstrut_1 a_{pq}$ respectively. This proves the lemma. ∎

Lemma 4. *If $h(p) \leq s, h(q) \leq s, \ell(p) \leq t, \ell(q) \leq t$, then*

(22) $$\mathstrut_1 b_{pq} = \mathstrut_1 b_{p_{s,t}^*, q_{s,t}^*}.$$

Proof. Let $A = \operatorname{diag}(\alpha_1, \ldots, \alpha_t)$, $q = (q_1, \ldots, q_\ell)$, $q_1 \leq s$, $\ell \leq t$. Then

(23) $$|A|^s M_q(A^{-1}) = M_{q_{s,t}^*}(A).$$

See the proof of Lemma 4.3.2. (Again (23) dos not hold for AM_q). Now (4.4.29), (4.4.31), (4.4.32) hold with $M_p, M_q, \mathstrut_1 b_{pq}$ replacing $U_p, U_q, \mathstrut_1 a_{pq}$ respectively. This proves the lemma. ∎

We have shown that the recurrence relations of Section 4.4.2 hold in exactly the same way for the coefficients of U_q's as for the coefficients of M_q's.

In the next section we discuss James' partial differential equation and a recurrence relation derived from it. The mathematical development will be somewhat sketchy.

4.5.4 James' partial differential equation and recurrence relation

James (1968) derived a partial differential equation satisfied by a zonal polynomial from the fact that a zonal polynomial is an "eigenfunction of the Laplace-Beltrami operator." Let $A=\text{diag}(\alpha_1, \ldots, \alpha_k)$, $p=(p_1, \ldots, p_\ell) \in P_n$. Then his partial differential equation is

$$(24) \quad \sum_{i=1}^{k} \alpha_i^2 \frac{\partial^2}{\partial \alpha_i^2} Y_p(A) + \sum_{i \neq j}^{k} \frac{\alpha_i^2}{\alpha_i - \alpha_j} \frac{\partial}{\partial \alpha_i} Y_p(A) = \left(\sum_{i=1}^{\ell} p_i(p_i - i + k - 1) \right) Y_p(A).$$

This might seem a little bit strange because it depends on the number of variables (k appears in the summation on the right hand side). Let

$$(25) \quad a_1(p) = \sum_{i=1}^{\ell} p_i(p_i - i).$$

Then the right hand side of (24) can be written as

$$(26) \quad a_1(p) Y_p(A) + n(k-1) Y_p(A), \qquad n = |p|.$$

To get rid of $n(k-1)Y_p(A)$ we notice the fact that for any $f \in V_n$, $\sum_{i=1}^{k} \alpha_i (\partial/\partial \alpha_i) f = nf$. Therefore

$$(27) \quad (k-1)n Y_p(A) = (k-1) \sum_{i=1}^{k} \alpha_i \frac{\partial}{\partial \alpha_i} Y_p(A).$$

But we can write

$$(28) \quad (k-1) \sum_{i=1}^{k} \alpha_i \frac{\partial}{\partial \alpha_i} Y_p(A) = \sum_{j=1}^{k} \sum_{i \neq j} \alpha_i \frac{\partial}{\partial \alpha_i} Y_p(A).$$

Now subtracting (28) from both sides of (24) and using the relation $\alpha_i^2/(\alpha_i - \alpha_j) - \alpha_i = \alpha_i \alpha_j/(\alpha_i - \alpha_j)$ we obtain

$$(29) \quad \sum_{i=1}^{k} \alpha_i^2 \frac{\partial^2}{\partial \alpha_i^2} Y_p(A) + \sum_{i \neq j}^{k} \frac{\alpha_i \alpha_j}{\alpha_i - \alpha_j} \frac{\partial}{\partial \alpha_i} Y_p(A) = a_1(p) Y_p(A),$$

which does not involve k as a coefficient and is valid for any number of variables. (29) was derived by Sugiura (1973) in an elementary way. Because his exposition is clear and readable (except that there are complications like a higher order partial differential equation and differential equations for complex zonal polynomials) we do not derive it here. Let

$$\mathcal{Y}_p(A) = \sum_{q \leq p} b_{pq} M_q(A). \tag{30}$$

Substituting this into (29) we obtain

$$\sum_{q \leq p} b_{pq} \left\{ \sum_{i=1}^{k} \alpha_i^2 \frac{\partial^2}{\partial \alpha_i^2} + \sum_{i \neq j}^{k} \frac{\alpha_i \alpha_j}{\alpha_i - \alpha_j} \frac{\partial}{\partial \alpha_i} \right\} M_q(A) = a_1(p) \sum_{q \leq p} b_{pq} M_q(A). \tag{31}$$

Now

$$\sum_{i=1}^{k} \alpha_i^2 \frac{\partial^2}{\partial \alpha_i^2} M_q(A), \qquad \sum_{i \neq j}^{k} \frac{\alpha_i \alpha_j}{\alpha_i - \alpha_j} \frac{\partial}{\partial \alpha_i} M_q(A) \tag{32}$$

can be expressed as sums of monomial symmetric functions. Then comparing both sides of (31) we can determine the coefficients b_{pq}. It is hard to visualize what is going on here unless one works out some examples. Muirhead (1982) does that very carefully using (24) rather than (29) following James (1968). Therefore we only sketch the procedure here.

Let $q = (q_1, \ldots, q_\ell)$. It is fairly straightforward to verify that

$$\sum_{i=1}^{k} \alpha_i^2 \frac{\partial^2}{\partial \alpha_i^2} M_q = \sum_{i=1}^{\ell} q_i(q_i - 1) M_q, \tag{33}$$

$$\sum_{i \neq j} \frac{\alpha_i \alpha_j}{\alpha_i - \alpha_j} \frac{\partial}{\partial \alpha_i} M_q = -\sum_{i=1}^{\ell} q_i(i-1) M_q + \text{lower order terms}. \tag{34}$$

§ 4.5 Coefficients of M_q

Adding (33) and (34) we obtain

(35) $$DM_q = a_1(q)M_q + \text{lower order terms},$$

where

$$D = \sum_{i=1}^{k} \alpha_i^2 \frac{\partial^2}{\partial \alpha_i^2} + \sum_{i \neq j}^{k} \frac{\alpha_i \alpha_j}{\alpha_i - \alpha_j} \frac{\partial}{\partial \alpha_i}.$$

It is fortunate to get only lower order terms by the differential operation. It is this triangular nature of the differential operation that enables one to determine b_{pq} recursively starting from the (arbitrary) leading coefficient b_{pp}. If one works out "lower order terms" (which is not hard) one arrives at the following rule by James (1968):

(36) $$b_{pq} = \sum_{q < q' \leq p} \frac{((q_i + r) - (q_j - r))b_{pq'}}{a_1(p) - a_1(q)},$$

where q' is an unordered partition of the form $q'=(q_1,\ldots,q_i + r,\ldots,q_j - r,\ldots,q_{\ell(q)})$, $(1 \leq r \leq q_j)$. The summation is over all (i,j,r) where $i < j$, $r \geq 1$ such that when the unordered partition q' is ordered we have $q < q' \leq p$.

Actually in view of Theorem 4.1.1 we have to consider only partitions q, q' which are majorized by p.

The advantage of this method is that it is self-contained. It gives all coefficients of a single zonal polynomial without computing others. Therefore it is by far the best method if one is interested in computing few zonal polynomials. On the other hand if one wants to compute many zonal polynomials then relying exclusively on this method seems to involve a great deal of redundant computations in view of recurrence relations of Section 4.5.3.

Remark 1. Logically the recurrence relation (36) is not complete until one shows that the denominator $a_1(p) - a_1(q)$ is never zero. James (1968) states that (36) gives rise to positive b_{pq}'s. Since the numerator $(q_i+r) - (q_j-r)=(q_i - q_j) + 2r$ is positive he seems to claim that $a_1(p) - a_1(q) > 0$ for all relevant pairs p, q. By Theorem 4.1.1 it is enough to prove

(37) $$a_1(p) - a_1(q) > 0 \quad \text{for } p \succ q.$$

Then this ensures that (36) works for all cases and nonzero b_{pq}'s are positive. (36) can be easily proved using techniques from the theory of majorization. See Marshall and Olkin (1979). We do not go into this here.

§ 4.6 COEFFICIENTS OF T_q IN Z_p

In this section we study the coefficients of T_p. The normalization Z_p seems to be most advantageous. An important fact about the coefficients of T_p is their orthogonality. We derive this first. See formulas (117) and (118) in James (1964), and Problem 13.3.9 in Farrell (1976).

For $p \in P_n$ let

$$(1) \qquad Z_p = \sum_{q \in P_n} g_{pq} T_q.$$

Let $\boldsymbol{Z} = (Z_{(n)}, Z_{(n-1,1)}, \ldots, Z_{(1^n)})'$, $\boldsymbol{G} = (g_{pq})$. Then (1) can be expressed in a matrix form as

$$(2) \qquad \boldsymbol{Z} = \boldsymbol{GT}.$$

Now we recall that the transition matrix $\boldsymbol{F}$ in $\boldsymbol{T} = \boldsymbol{FU}$ is lower triangular (see (2.2.28)). Substituting this into (2) we obtain

$$(3) \qquad \boldsymbol{Z} = \boldsymbol{GFU}.$$

But $\boldsymbol{Z} = \boldsymbol{BU}$. Hence

$$(4) \qquad \boldsymbol{G} = \boldsymbol{BF}^{-1}$$

Now (3.4.11) shows

$$(5) \qquad \boldsymbol{B'DB} = \boldsymbol{F'CF},$$

where $\boldsymbol{C} = \mathrm{diag}(c_p, p \in P_n)$ is obtained in (3.4.8) and $\boldsymbol{D} = \mathrm{diag}(d_p, p \in P_n)$ is known as (3.4.12). From (4) and (5) we obtain

$$(6) \qquad \boldsymbol{G'DG} = \boldsymbol{C}.$$

Inverting this

(7) $$GC^{-1}G' = D^{-1}.$$

Coordinatewise

(8) $$\sum_p d_p g_{pq} g_{pq'} = \delta_{qq'} c_q, \quad \text{(column orthogonality)},$$

(9) $$\sum_q g_{pq} g_{p'q} / c_q = \delta_{pp'} / d_p, \quad \text{(row orthogonality)},$$

where $\delta_{pp'}$ is Kronecker's delta.

Actually $c_q, q \in P_n$ coincide with the elements of the first row of G. To see this let $\beta_1 = 1, \beta_2 = \cdots = \beta_k = 0$ in (3.4.7). Then clearly $T_p(I_1) = 1$ for every p and we have

(10) $$\prod_{i=1}^{k} (1 - 2\theta\alpha_i)^{-1/2} = \sum_{n=0}^{\infty} (\theta^n / n!) \sum_{p \in P_n} c_p T_p(A).$$

Comparing this to (4.4.4) we obtain $Z_{(n)}(A) = \sum c_p T_p(A)$ and hence $c_p = g_{(n),p}$. Therefore (9) can be written alternatively as

(11) $$\sum_q g_{pq} g_{p'q} / g_{(n),q} = \delta_{pp'} / d_p.$$

One obvious advantage of working with T_p's is that the coefficient matrix is readily invertible. From (6)

(12) $$G^{-1} = C^{-1} G' D.$$

Therefore once we express zonal polynomials in terms of T_p's then it is easy to express T_p's (and their linear combinations) in zonal polynomials.

(11) was used to compute zonal polynomials in Parkhurst and James (1974) as follows. (i) U_p's are expressed in T_p's. (ii) They are Gram-Schmidt orthogonalized relative to the orthogonality relation (11) starting from the lowest

partition (1^n) upwards. Because of the triangularity of B this clearly results in zonal polynomials.

Some g_{pq}'s can be explicitly obtained using the fact $Z_p(I_k) = \lambda_{kp}$. We regard λ_{kp} as a function in k. Then by (3.1.15) it is a polynomial in k of degree $|p| = n$. Now since $t_r(I_k) = k$ for any r we obtain $\mathcal{T}_p(I_k) = k^{p_1-p_2} k^{p_2-p_3} \ldots = k^{p_1} = k^{h(p)}$. Therefore putting I_k in (1) we obtain

$$(13) \qquad \lambda_{kp} = \sum_{q \in \mathcal{P}_n} g_{pq} k^{h(q)}.$$

This uniquely determines g_{pq} for $q = (n)$, $q = (n-1,1)$, $q = (1^n)$ because these are the only partitions in $\mathcal{P}_n$ with $h(q) = n, n-1, 1$ respectively. Now the coefficient of k^n in λ_{kp} is 1, hence

$$(14) \qquad g_{p,(n)} = 1.$$

(14) was originally used by James to determine the normalization Z_p. Now let us look at the coefficient of k^{n-1} in λ_{kp}. It is

$$(15) \qquad \begin{aligned} &\sum_{i=1}^{\ell(p)} \{(-i+1) + (-i+3) + \cdots + (-i+1+2p_i-2)\} \\ &= \sum_{i=1}^{\ell(p)} \frac{1}{2} p_i \{(-i+1) + (-i+1+2p_i-2)\} \\ &= \sum_{i=1}^{\ell(p)} p_i(p_i - i) \\ &= a_1(p). \end{aligned}$$

$a_1(p)$ already appeared in (4.5.25). Therefore

$$(16) \qquad g_{p,(n-1,1)} = a_1(p).$$

This is mentioned in the introductory part of Parkhurst and James (1964) in a somewhat different form

$$(17) \qquad g_{p,(n-1,1)} = \sum_{i=1}^{\ell(p)} p_i(p_i - 1) - \frac{1}{2} \sum_{j=1}^{h(p)} p'_j(p'_j - 1),$$

where $p' = (p'_1, \ldots, p'_{h(p)})$ is the conjugate partition of p. Using (2.1.4) it is easy to check that (16) and (17) are equivalent.

Now the coefficient of k in λ_{kp} is

(18)
$$\{2 \cdot 4 \cdots (2p_1 - 2)\}\{(-1)(1)\cdots(2p_2 - 3)\}\cdots = 2^{n-1}(p_1 - 1)! \prod_{i=2}^{\ell(p)} \left(-\frac{i-1}{2}\right)_{p_i}.$$

Hence

(19)
$$g_{p,(1^n)} = 2^{n-1}(p_1 - 1)! \prod_{i=2}^{\ell(p)} \left(-\frac{i-1}{2}\right)_{p_i}.$$

This does not seem to have been noticed.

Now from (4)

(20)
$$\mathcal{B} = \mathbf{GF}, \quad \mathcal{B} = (\xi_{pq}), \quad \mathbf{F} = (f_{pq}).$$

$\mathbf{F}$ is lower triangular. Therefore the last column of $\mathcal{B}$ is $f_{(1^n)(1^n)}$ times the last column of $\mathbf{G}$. By (2.2.22)

(21)
$$\mathcal{T}_{(1^n)} = t_n = (-1)^{n-1} n (\mathcal{U}_{(1^n)} + \cdots).$$

Hence $f_{(1^n)(1^n)} = (-1)^{n-1} n$. Therefore

(22)
$$\xi_{p,(1^n)} = f_{(1^n)(1^n)} g_{p,(1^n)} = (-2)^{n-1} n (p_1 - 1)! \prod_{i=2}^{\ell(p)} \left(-\frac{i-1}{2}\right)_{p_i}.$$

Making use of (12) gives another set of identities.

(23)
$$\mathcal{T} = \mathbf{G}^{-1} \mathbf{Z} = \mathbf{C}^{-1} \mathbf{G}' \mathbf{D} \mathbf{Z}.$$

Now $\mathbf{DZ} = (d_{(n)} Z_{(n)}, \ldots, d_{(1^n)} Z_{(1^n)})'$ and $d_p Z_p$ is denoted by C_p (3.4.13). Therefore

(24)
$$\mathcal{T} = \mathbf{C}^{-1} \mathbf{G}' (C_{(n)}, \ldots, C_{(1^n)})'.$$

Comparing the second element we obtain

$$T_{(n-1,1)} = t_1^{n-2} t_2$$

(25)
$$= \frac{1}{c_{(n-1,1)}} \sum_{p \in \mathcal{P}_n} a_1(p) C_p$$

$$= \frac{1}{n(n-1)} \sum_{p \in \mathcal{P}_n} a_1(p) C_p,$$

where $c_{(n-1,1)} = n(n-1)$ is given by (3.4.8). (25) was given by Sugiura and Fujikoshi (1959) by a different method. They derive more identities of this kind. See Sugiura (1971) too. Now looking at the last element we obtain

$$T_{(1^n)} = t_n = M_{(n)}$$

(26)
$$= \frac{1}{c_{(1^n)}} 2^{n-1} \sum_{p \in \mathcal{P}_n} C_p \cdot (p_1 - 1)! \prod_{i=2}^{\ell(p)} \left(-\frac{i-1}{2} \right)_{p_i}$$

$$= \frac{1}{(n-1)!} \sum_{p \in \mathcal{P}_n} C_p \cdot (p_1 - 1)! \prod_{i=2}^{\ell(p)} \left(-\frac{i-1}{2} \right)_{p_i}.$$

What are advantages and disadvantages of working with T_p's? One advantage is that we do not have to compute characteristic roots of $\boldsymbol{A}$ to compute $T_p(\boldsymbol{A})$. (One only needs traces of powers of $\boldsymbol{A}$.) Another advantage is the orthogonality discussed above. A serious drawback of T_p is that we have to compute $T_p(\boldsymbol{A})$ for all $p \in \mathcal{P}_n$ even if the rank of $\boldsymbol{A}$ is small. In usual statistical computations rank $\boldsymbol{A}$ is fixed and not too large. It is a covariance matrix for example. Since the number of partitions grows very fast as n increases if one wants to compute $Z_p(\boldsymbol{A})$ for $|p|$ large it seems better to use U_q's or M_q's. The growth of the number of partitions p with $\ell(p) \leq k$ (k : fixed) is much smaller than the growth of the number of all partitions. See Table 4.1 in David, Kendall, and Barton (1966).

§ 4.7 VARIATIONS OF THE INTEGRAL REPRESENTATION OF ZONAL POLYNOMIALS

In this section we explore various variations of the integral representation (Theorem 3.3.1) discussed in Section 3.3. We first replace $\boldsymbol{U}$ by the $k \times k$ uniform orthogonal matrix $\boldsymbol{H}$.

§ 4.7 Variations of the integral representation

Theorem 1. *(James, 1973)*   For $k \times k$ symmetric A

$$
\begin{aligned}
(1) \quad \frac{\mathcal{Y}_p(A)}{\mathcal{Y}_p(I_k)} &= \mathcal{E}_H\{\Delta_1^{p_1-p_2} \cdots \Delta_\ell^{p_\ell}\} \\
&= \mathcal{E}_H\{\prod_{i=1}^{\ell} [HAH'(1,\ldots,i)]^{p_i-p_{i+1}}\},
\end{aligned}
$$

where $p = (p_1,\ldots,p_\ell) \in P_n$, $k \geq \ell$, and the $k \times k$ orthogonal H is uniformly distributed.

Proof.   As in Lemma 3.1.2 it is easy to check that $\mathcal{E}_H\{\Delta_1^{p_1-p_2} \cdots \Delta_\ell^{p_\ell}\} \in V_{n,k}$. Therefore we can write

$$
(2) \quad \mathcal{E}_H\{\Delta_1^{p_1-p_2} \cdots \Delta_\ell^{p_\ell}\} = \sum_{q \in P_n, \ell(q) \leq k} a_q Z_q(A).
$$

Replacing A by UAU' where $k \times k$ U is as in Theorem 3.3.1 and taking expectation with respect to U we obtain

$$
(3) \quad Z_p(A) = \sum_{q \in P_n, \ell(q) \leq k} a_q \lambda_{kq} Z_q(A).
$$

This being true for any symmetric $k \times k$ A we conclude from Theorem 4.1.1

$$
a_q = 0,\ q \neq p,\ a_p = \frac{1}{\lambda_{kp}} = \frac{1}{Z_p(I_k)}.
$$

Since (1) is independent of normalization we can have $\mathcal{Y}_p$ instead of Z_p in (1). ∎

Corollary 1.   *(Kates)*   Let $X : k \times k$ have an orthogonally biinvariant distribution then

$$
(4) \quad \mathcal{E}_X\{\Delta_1^{p_1-p_2} \cdots \Delta_\ell^{p_\ell}\} = \frac{\mathcal{Y}_p(A)\mathcal{E}_X\{\mathcal{Y}_p(X'X)\}}{\{\mathcal{Y}_p(I_k)\}^2}.
$$

where $\Delta_i = XAX'(1,\ldots,i)$ and A is symmetric.

Proof. We replace X by $H_1 X H_2$ where H_1 and H_2 are independently uniformly distributed. The distribution of X is unchanged. Now taking expectation with respect to H_1 (Theorem 1) and H_2 (Theorem 3.2.1) successively we obtain

(5)
$$\mathcal{E}_X\{\Delta_1^{p_1-p_2}\cdots\Delta_\ell^{p_\ell}\} = \mathcal{E}_{H_1,X,H_2}\{\prod_{i=1}^{\ell}[H_1 X H_2 A H_2' X' H_1'(1,\ldots,i)]^{p_i-p_{i+1}}\}$$
$$= \mathcal{E}_{X,H_2}\{\mathcal{Y}_p(X H_2 A H_2' X')\}/\mathcal{Y}_p(I_k)$$
$$= \mathcal{Y}_p(A)\mathcal{E}_X\{\mathcal{Y}_p(X'X)\}/\{\mathcal{Y}_p(I_k)\}^2.$$

∎

Remark 1. As in Remark 3.2.5 X can be rectangular. If X is $m \times k$, then $\{\mathcal{Y}_p(I_k)\}^2$ on the right hand side of (4) is replaced by $\mathcal{Y}_p(I_k)\mathcal{Y}_p(I_m)$.

An easy modification of the above formulas produces another set of identities.

Theorem 2. *Let U_1, U_2 be $k \times k$ matrices whose entries are independent standard normal variables. Then for $k \times k$ A*

(6) $$_1 b_p Z_p(AA') = \mathcal{E}_{U_1,U_2}\{\prod_{i=1}^{\ell}[U_1 A U_2(1,\ldots,i)]^{2p_i-2p_{i+1}}\},$$

where $_1 b_p$ is given by (4.2.7).

Proof. Let the singular value decomposition of A be $A = \Gamma_1 D \Gamma_2$ where Γ_1, Γ_2 are orthogonal, $D = \text{diag}(\delta_1,\ldots,\delta_k)$ and $\delta_1^2, \ldots, \delta_k^2$ are the characteristic roots of AA'. Since the order of $\delta_1, \ldots, \delta_k$ and the sign of each δ_i can be arbitrary in the singular value decomposition we see that (6) is a homogeneous symmetric polynomial in $\delta_1^2, \ldots, \delta_k^2$. Denote the right hand side of (6) by $f(AA')$. We use the converse part of Theorem 3.1.1. We want to show that if $W \sim \mathcal{W}(I_k, \nu)$ then

(7) $$\mathcal{E}_W f(AA'W) = \lambda_{\nu p} f(AA')$$

for all sufficiently large ν. Now fix ν and let A, U_1, U_2 be augmented to $\nu \times \nu$ as in the proofs of Theorem 3.2.4 or Theorem 3.3.1. (Here we do not place $\sim$

§ 4.7　　　*Variations of the integral representation*　　　81

for notational simplicity.) Note that $f(AA')$ does not change by this change of dimensionality. Also in (7) W can be augmented to be $\nu \times \nu$ by considering W as $k \times k$ upper left submatrix of a Wishart matrix $\mathcal{W}(I_\nu,\nu)$. Now let U_3 ($\nu \times \nu$) be distributed independently of U_1 and U_2. Then

(8)
$$\begin{aligned}
\mathcal{E}_W f(AA'W) &= \mathcal{E}_{U_3} f(AA'U_3 U_3') = \mathcal{E}_{U_3} f(U_3 AA' U_3') \\
&= \mathcal{E}_{U_1,U_2,U_3}\{ \prod_{i=1}^{\ell} [U_1 U_3 A U_2(1,\ldots,i)]^{2p_i - 2p_{i+1}} \} \\
&= \mathcal{E}_{U_1,U_2,U_3}\{ \prod_{i=1}^{\ell} [U_3 U_1 A U_2(1,\ldots,i)]^{2p_i - 2p_{i+1}} \} \\
&= \mathcal{E}_{U_1,U_2,T,H}\{ \prod_{i=1}^{\ell} [THU_1 A U_2(1,\ldots,i)]^{2p_i - 2p_{i+1}} \} \\
&= \lambda_{\nu p} \mathcal{E}_{U_1,U_2}\{ \prod_{i=1}^{\ell} [U_1 A U_2(1,\ldots,i)]^{2p_i - 2p_{i+1}} \} \\
&= \lambda_{\nu p} f(AA'),
\end{aligned}$$

where T, H are as in Lemma 3.2.2. Therefore by Theorem 3.1.1 $f = c Z_p$ for some c. To obtain c we put $A = I_\nu$. Then

(9)
$$\begin{aligned}
c\lambda_{\nu p} &= \mathcal{E}_{U_1,U_2}\{ \prod_{i=1}^{\ell} [U_1 U_2(1,\ldots,i)]^{2p_i - 2p_{i+1}} \} \\
&= \mathcal{E}_{T,U_2}\{ \prod_{i=1}^{\ell} [T U_2(1,\ldots,i)]^{2p_i - 2p_{i+1}} \} \\
&= \lambda_{\nu p} \mathcal{E}_{U_2}\{ \prod_{i=1}^{\ell} U_2(1,\ldots,i)^{2p_i - 2p_{i+1}} \}.
\end{aligned}$$

Hence $c = \mathcal{E}_U\{\prod_{i=1}^{\ell} U(1,\ldots,i)^{2p_i-2p_{i+1}}\}$. Now consider (3.3.2) and (3.3.8). Then we see that the coefficient of $\alpha_1^{p_1}\cdots\alpha_\ell^{p_\ell}$ is given just by $\mathcal{E}_U\{\prod_{i=1}^{\ell} U(1,\ldots,i)^{2p_i-2p_{i+1}}\}$. Therefore it is the leading coefficient of Z_p and is equal to $_1 b_p$ given in (4.2.7). ∎

Corollary 2.

(10)
$$\frac{_1 b_p Z_p(AA')}{Z_p(I_k)} = \mathcal{E}_{H_1,U_2}\{ \prod_{i=1}^{\ell} [H_1 A U_2(1,\ldots,i)]^{2p_i - 2p_{i+1}} \}.$$

(11) $$\frac{1^{b_p} Z_p(AA')}{\{Z_p(I_k)\}^2} = \mathcal{E}_{H_1,H_2}\{\prod_{i=1}^{\ell}[H_1AH_2(1,\ldots,i)]^{2p_i-2p_{i+1}}\}.$$

Proof. (10) and (11) can be proved successively as in the proof of Theorem 2. ∎

Corollary 3. *Let X_1, X_2 be independent and have orthogonally biinvariant distributions. Then*

(12) $$Z_p(AA')\frac{1^{b_p}\mathcal{E}_{X_1}\{Z_p(X_1X_1')\}\mathcal{E}_{X_2}\{Z_p(X_2X_2')\}}{\{Z_p(I_k)\}^4}$$
$$= \mathcal{E}_{X_1,X_2}\{\prod_{i=1}^{\ell}[X_1AX_2(1,\ldots,i)]^{2p_i-2p_{i+1}}\}.$$

Proof. Replace X_1 by $H_1X_1H_3$ and X_2 by $H_4X_2H_2$. Then taking expectation with respect to H_1, H_2, H_3, H_4 successively we obtain (12). ∎

Remark 2. Generalization to rectangular matrices is straightforward.

CHAPTER 5

Complex zonal polynomials

In this chapter we study complex zonal polynomials, i.e. zonal polynomials associated with the complex normal and the complex Wishart distributions. The complex multivariate normal distribution is used in the frequency analysis of multiple time series and complex zonal polynomials are useful for noncentral distributions arising in this setting. Other than that the practical applicability of complex zonal polynomials seems rather limited. Actually our main reason of studying them is that they are *simpler* than real zonal polynomials. If one compares Farrell (1980) and Chapter 1 of Macdonald (1979) it becomes apparent that complex zonal polynomials are the same as homogeneous symmetric polynomials called the Schur functions and the latter have been extensively studied. We will make this connection clear. Hopefully developing complex zonal polynomials gives further insights into the real case.

The theory of the complex normal and the Wishart distributions very closely parallels that of the real case (see Goodman (1963) or Brillinger (1975)) and it turns out that our development of Chapter 3 and Chapter 4 can be directly translated into the complex case. In the literature on zonal polynomials it seems customary to put a $\sim$ to denote corresponding objects in the complex case. For example we use $\tilde{Z}_p, \tilde{C}_p, \tilde{Y}_p$, etc. With this convention the translation of the results in Chapter 3 and 4 are almost immediate.

§ 5.1 THE COMPLEX NORMAL AND THE COMPLEX WISHART DISTRIBUTIONS

We give a brief summary of the complex normal and the complex Wishart distributions. Let x, y be independently distributed according to $N(0, 1/2)$ and let $z = x + iy$. z is said to have *the standard complex normal distribution*. Or we say that z is a *standard complex normal (random) variable*. Now let A be an $n \times n$ matrix with complex elements and let

$$(1) \qquad \boldsymbol{u} = (u_1, \ldots, u_k)' = \boldsymbol{A}\, (z_1, \ldots, z_k)',$$

where $z_1, \ldots, z_k$ are independent standard complex normal variables. This scheme generates a family of distributions called the multivariate complex normal distribution. Its density (with respect to $\prod_1^k d(\Re u_i)\ \prod_1^k d(\Im u_i)$) is given by

$$(2) \qquad f(\boldsymbol{u}) = \frac{1}{\pi^k |\Sigma|} \exp(-\boldsymbol{u}^* \Sigma^{-1} \boldsymbol{u})$$

where $*$ means conjugate transpose and $\Sigma = \mathcal{E}\boldsymbol{u}\boldsymbol{u}^* = \boldsymbol{A}\boldsymbol{A}^*$. If $\boldsymbol{u}$ has the density (2) we denote this by $\boldsymbol{u} \sim CN(\boldsymbol{0}, \Sigma)$. Now suppose that $\boldsymbol{u}_1, \ldots, \boldsymbol{u}_n$ are independently distributed according to $CN(\boldsymbol{0}, \Sigma)$. Let $\tilde{\boldsymbol{W}} = \sum_{i=1}^n \boldsymbol{u}_i \boldsymbol{u}_i^*$. The distribution of $\tilde{\boldsymbol{W}}$ is called the complex Wishart distribution and its density (with respect to $\prod_{i=1}^k d\tilde{w}_{ii}\ \prod_{i<j} d(\Re \tilde{w}_{ij}) d(\Im \tilde{w}_{ij})$) is given by

$$(3) \qquad f(\tilde{\boldsymbol{w}}) = \frac{|\tilde{\boldsymbol{w}}|^{n-k} \exp(-\operatorname{tr} \Sigma^{-1} \tilde{\boldsymbol{w}})}{\pi^{p(p-1)/2} \prod_{i=1}^k \Gamma(n-i+1) |\Sigma|^n}$$

This distribution is denoted by $CW(\Sigma, n)$.

Let $\tilde{\boldsymbol{W}} = \tilde{\boldsymbol{T}}\tilde{\boldsymbol{T}}^*$ be the (unique) triangular decomposition of a positive definite Hermitian matrix where $\tilde{\boldsymbol{T}} = (\tilde{t}_{ij})$ is a lower triangular matrix with positive diagonal elements. Analogous to Lemma 3.1.3 we have the following lemma.

Lemma 1. *Let $\tilde{\boldsymbol{W}}$ be distributed according to $CW(I_k, \nu)$. Let $\tilde{\boldsymbol{W}} = \tilde{\boldsymbol{T}}\tilde{\boldsymbol{T}}^*$. Then $\tilde{t}_{ij}, i \geq j$, are independently distributed. $2^{1/2}\,\tilde{t}_{ii} \sim \chi(2(\nu - i + 1))$ and $\tilde{t}_{ij}, i > j$, are standard complex normal variables.*

Proof.   See Goodman (1963), formula (1.8). ∎

§ 5.2   Derivation of complex zonal polynomials

Remark 1. $\tilde{t}_{ii}^2$ has the gamma density $f(x) = (1/\Gamma(\nu - i + 1))x^{\nu-i}e^{-x}$.

With this lemma we are ready to translate the results of Chapter 3 and 4.

§ 5.2 DERIVATION AND PROPERTIES OF COMPLEX ZONAL POLYNOMIALS

For ease of comparison of the results here and the results of Chapter 3 and 4 we will consistently put $\sim$ on corresponding objects of the complex case. This sometimes results in somewhat unnatural notation, for example if H is orthogonal then $\tilde{H}$ is unitary etc. So much for the notation; now let us follow the development of real zonal polynomials step by step for a while. All proofs will be omitted since they are the same for the real and the complex cases.

We consider the following transformation.

(1) $$(\tilde{\tau}_\nu \, \mathcal{U}_p)(\tilde{A}) = \mathcal{E}_{\tilde{W}}\{\mathcal{U}_p(\tilde{A}\tilde{W})\},$$

where $\tilde{A}$ is Hermitian and $\tilde{W} \sim C\mathcal{W}(I_k, \nu)$.

Lemma 1.   *(corresponding to Lemma 3.1.2)* $\tilde{\tau}_\nu \, \mathcal{U}_p \in V_n$.

Corollary 1.   *(Corollary 3.1.1)*

(2) $$(\tilde{\tau}_\nu \, \mathcal{U}_p)(\tilde{A}) = \tilde{\lambda}_{\nu p} \, \mathcal{U}_p(\tilde{A}) + \sum_{q<p} \tilde{a}_{pq} \, \mathcal{U}_q(\tilde{A}).$$

Corollary 2.   *(Corollary 3.1.2)*

(3)
$$\begin{aligned}
\tilde{\lambda}_{\nu p} &= \prod_{i=1}^{\ell(p)} \frac{\Gamma(p_i + \nu + 1 - i)}{\Gamma(\nu + 1 - i)} \\
&= \prod_{i=1}^{\ell(p)} (\nu + 1 - i)_{p_i} \\
&= \nu(\nu + 1) \cdots (\nu + p_1 - 1) \\
&\quad \cdot (\nu - 1)\nu \cdots (\nu - 1 + p_2 - 1) \\
&\quad \cdots \\
&\quad \cdot (\nu - \ell + 1) \cdots (\nu - \ell + p_\ell)
\end{aligned}$$

where $\ell = \ell(p)$ and $(a)_k = a(a+1)\cdots(a+k-1)$.

Corollary 1 shows that

$$\tilde{\tau}_\nu(\mathcal{U}) = \tilde{T}_\nu \mathcal{U}, \tag{4}$$

where $\tilde{T}_\nu$ is an upper triangular matrix with diagonal elements $\tilde{t}_{pp} = \tilde{\lambda}_{\nu p}$.

Lemma 2. *(Lemma 3.1.4) There exists a nonsingular upper triangular matrix $\tilde{B}$ such that*

$$\tilde{B}\tilde{T}_\nu = \tilde{\Lambda}_\nu \tilde{B} \quad \text{for all } \nu, \tag{5}$$

where $\tilde{\Lambda}_\nu = \mathrm{diag}(\tilde{\lambda}_{\nu p}, p \in \mathcal{P}_n)$. $\tilde{B}$ is uniquely determined up to a (possibly different) multiplicative constant for each row.

Using this $\tilde{B}$ we define complex zonal polynomials.

Definition 1. *(Definition 3.1.1)*   *Complex zonal polynomials*

Let $\tilde{B}$ be as in Lemma 2. Complex zonal polynomials $\tilde{y}_p$, $p \in \mathcal{P}_n$ are defined by

$$\tilde{y} = \begin{pmatrix} \tilde{y}_{(n)} \\ \tilde{y}_{(n-1,1)} \\ \cdot \\ \cdot \\ \cdot \\ \tilde{y}_{(1^n)} \end{pmatrix} = \tilde{B}\mathcal{U}. \tag{6}$$

Lemma 2 is a consequence of the fact that there exists ν_0 for which $\tilde{\lambda}_{\nu_0 p}$, $p \in \mathcal{P}_n$ are all different and the following lemma.

Lemma 3. *(Lemma 3.1.5)*

$$\tilde{T}_\nu \tilde{T}_\mu = \tilde{T}_\mu \tilde{T}_\nu. \tag{7}$$

We summarize these results in the following theorem.

Theorem 1. *(Theorem 3.1.1) Let $\tilde{y}_p$ be a complex zonal polynomial then*

$$\mathcal{E}_{\tilde{W}} \tilde{y}_p(\tilde{A}\tilde{W}) = \tilde{\lambda}_{\nu p} \tilde{y}_p(\tilde{A}), \tag{8}$$

where $\tilde{W} \sim C\mathcal{W}(I_k, \nu)$, $\tilde{A}$ is Hermitian, and $\tilde{\lambda}_{\nu p}$ is given by (3). Conversely (8) (for all sufficiently large ν and for all Hermitian $\tilde{A}$) implies that $\tilde{y}_p$ is a complex zonal polynomial.

Now we explore various integral identities satisfied by complex zonal polynomials. The uniform distribution of unitary matrices can be defined as in the case of orthogonal matrices. In particular we have

Lemma 4. *(Lemma 3.2.2) Let $\tilde{U} = (\tilde{u}_{ij})$ be a $k \times k$ matrix such that $\tilde{u}_{ij}$ are independent standard complex normal variables. Then with probability 1 $\tilde{U}$ can be uniquely expressed as*

$$\tilde{U} = \tilde{T}\tilde{H}, \tag{9}$$

where $\tilde{T} = (\tilde{t}_{ij})$ is lower triangular with positive diagonal elements and $\tilde{H}$ is unitary. Furthermore (i) $\tilde{T}, \tilde{H}$ are independent, (ii) $\tilde{H}$ is uniform, (iii) $\tilde{t}_{ij}$ are all independent and $2^{1/2}\tilde{t}_{ii} \sim \chi(2(k-i+1))$, $\tilde{t}_{ij}$, $i > j$, $\sim C\mathcal{N}(0,1)$.

Now we obtain the "splitting property" of complex zonal polynomials.

Theorem 2. *(Theorem 3.2.1) Let $\tilde{A}, \tilde{B}$ be $k \times k$ Hermitian matrices. Then*

$$\mathcal{E}_{\tilde{H}} \tilde{y}_p(\tilde{A}\tilde{H}\tilde{B}\tilde{H}^*) = \tilde{y}_p(\tilde{A})\tilde{y}_p(\tilde{B})/\tilde{y}_p(I_k), \tag{10}$$

where $k \times k$ unitary $\tilde{H}$ has the uniform distribution.

Definition 2. A random Hermitian matrix $\tilde{V}$ is said to have a *unitarily invariant distribution* if for every unitary $\tilde{\Gamma}$, $\tilde{\Gamma}\tilde{V}\tilde{\Gamma}^*$ has the same distribution as $\tilde{V}$.

As in the real case Theorem 1 generalizes to unitarily invariant distributions.

Theorem 3. *(Theorem 3.2.2) Suppose that $\tilde{V}$ has a unitarily invariant distribution, then for Hermitian $\tilde{A}$*

$$\mathcal{E}_{\tilde{V}}\,\tilde{y}_p(\tilde{A}\tilde{V}) = c_p\,\tilde{y}_p(\tilde{A}), \tag{11}$$

where

$$c_p = \mathcal{E}_{\tilde{V}}\{\tilde{y}_p(\tilde{V})\}/\tilde{y}_p(I_k). \tag{12}$$

Unitarily invariant distributions are characterized as follows.

Lemma 5. *(Lemma 3.2.3) Let $\tilde{V} = \tilde{H}\tilde{D}\tilde{H}^*$ where $\tilde{H}$ is unitary and $\tilde{D}$ is diagonal. Let $\tilde{H}$ and $\tilde{D}$ be independently distributed such that $\tilde{H}$ has the uniform distribution. (Diagonal elements of $\tilde{D}$ can have any distribution.) Then $\tilde{V}$ has a unitarily invariant distribution. Conversely all unitarily invariant distributions can be obtained in this way.*

We can replace $\tilde{H}$ in Theorem 2 by $\tilde{U}$ whose elements are independent standard complex normal variables.

Theorem 4. *(Theorem 3.2.3) Let $\tilde{U} = (\tilde{u}_{ij})$ be a $k \times k$ matrix such that $\tilde{u}_{ij}$ are independent standard complex normal variables. Then for Hermitian $\tilde{A}, \tilde{B}$*

$$\mathcal{E}_{\tilde{U}}\,\tilde{y}_p(\tilde{A}\tilde{U}\tilde{B}\tilde{U}^*) = \frac{\tilde{\lambda}_{kp}}{\tilde{y}_p(I_k)}\tilde{y}_p(\tilde{A})\tilde{y}_p(\tilde{B}). \tag{13}$$

As in the real case this leads to the following observation.

Theorem 5. *(Theorem 3.2.4) $\tilde{b}_p \equiv \tilde{\lambda}_{kp}/\tilde{y}_p(I_k)$ is a constant independent of k.*

Unitarily biinvariant distributions are defined in an obvious way.

Definition 3. A random matrix $\tilde{X}$ has a *unitarily biinvariant distribution* if for every unitary $\tilde{\Gamma}_1, \tilde{\Gamma}_2$, the distribution of $\tilde{\Gamma}_1 \tilde{X} \tilde{\Gamma}_2$ is the same as the distribution of $\tilde{X}$.

Now Theorem 2 and Theorem 4 generalize as follows.

§ 5.2   Derivation of complex zonal polynomials

Theorem 6. *(Theorem 3.2.5)*   *If $\tilde{X}$ has a unitarily biinvariant distribution then for Hermitian $\tilde{A}, \tilde{B}$*

$$\mathcal{E}_{\tilde{X}} \tilde{\mathcal{Y}}_p(\tilde{A}\tilde{X}\tilde{B}\tilde{X}^*) = \gamma_p \, \tilde{\mathcal{Y}}_p(\tilde{A}) \, \tilde{\mathcal{Y}}_p(\tilde{B}), \tag{14}$$

where

$$\gamma_p = \mathcal{E}_{\tilde{X}}\{\tilde{\mathcal{Y}}_p(\tilde{X}\tilde{X}^*)\}/\{\tilde{\mathcal{Y}}_p(I_k)\}^2. \tag{15}$$

Characterization of unitarily biinvariant distributions can be given in an obvious way.

Lemma 6. *(Lemma 3.2.4)*   *Let $\tilde{X} = \tilde{H}_1 \tilde{D} \tilde{H}_2$ where $\tilde{H}_1, \tilde{H}_2$ are unitary and $\tilde{D}$ is diagonal. Let $\tilde{H}_1, \tilde{H}_2, \tilde{D}$ be independently distributed such that $\tilde{H}_1, \tilde{H}_2$ have the uniform distribution. ($\tilde{D}$ can have any distribution.) Then $\tilde{X}$ has a unitarily biinvariant distribution. Conversely all unitarily biinvariant distributions can be obtained in this way.*

Remark 1. *The notion of unitarily biinvariant distributions applies to rectangular matrices as well.*

Now we take a look at the integral representation of zonal polynomials in the complex case.

Definition 4. A particular normalization of a zonal polynomial denoted by $\tilde{Z}_p$ is defined by

$$\tilde{Z}_p(I_k) = \tilde{\lambda}_{kp}, \tag{16}$$

or $\tilde{b}_p = 1$ in Theorem 5.

Theorem 7. *(Theorem 3.3.1)*   *Let $p = (p_1, \ldots, p_\ell)$. For $k \times k$ Hermitian $\tilde{A}$*

$$\tilde{Z}_p(\tilde{A}) = \mathcal{E}_{\tilde{U}}\{\tilde{\Delta}_1^{p_1-p_2} \tilde{\Delta}_2^{p_2-p_3} \cdots \tilde{\Delta}_\ell^{p_\ell}\}, \tag{17}$$

where $\tilde{\Delta}_i = \tilde{U}\tilde{A}\tilde{U}^*(1,\ldots,i)$ *is the determinant of the* $i \times i$ *upper left minor of* $\tilde{U}\tilde{A}\tilde{U}^*$ *and* $\tilde{U}$ *is a* $k \times k$ *random matrix whose entries are independent standard complex normal variables.*

(17) implies that $\tilde{Z}_p(\tilde{A})$ is positive for positive definite $\tilde{A}$ and increasing in each root. Furthermore using the Gale-Ryser theorem (see Remark 4.1.1 and Remark 4.1.2) the coefficients of M_q in $\tilde{Z}_p$ are nonnegative and they are positive iff $p \succ q$.

As in the real case ${}_1\tilde{b}_p$ denotes the leading coefficient of $\tilde{Z}_p$, namely

(18) $$\tilde{Z}_p = {}_1\tilde{b}_p \, {}_1\tilde{y}_p.$$

Theorem 8. *(Theorem 4.2.2)*

(19) $$\begin{aligned}{}_1\tilde{b}_p &= \prod_{i=1}^{\ell(p)} \prod_{j=1}^{i} (i-j+1+p_j-p_i)_{p_i-p_{i+1}} \\ &= \frac{\prod_{i=1}^{\ell(p)}(p_i - i + \ell(p))!}{\prod_{i<j}(p_i - p_j - i + j)}.\end{aligned}$$

Other than mentioning Theorem 4.2.2 we will not follow the development of Chapter 4. Of course all the results of Chapter 4 can be translated into the complex case as has been done so far. However, it is pointless to go into numerical aspects of complex zonal polynomials because, as mentioned above, complex zonal polynomials are the Schur functions and the Schur functions are already well studied. Although the translation of the results in Chapter 4 presents an alternative "probabilistic" derivation of properties of the Schur functions, it is hardly more advantageous than a well developed standard approach to the subject. See Chapter 1 of Macdonald (1979) for example. The link between complex zonal polynomials and the Schur functions is given by Saw's generating function.

Saw's generating function in the complex case was introduced by Farrell (1980). Let $\tilde{u}_{ij}$ be a standard complex normal variable. Then $2|\tilde{u}_{ij}|^2 =$

§ 5.3   Schur functions   91

$2\tilde{u}_{ij}\tilde{u}_{ij}^* \sim \chi^2(2)$ (i.e. $|\tilde{u}_{ij}|^2$ has the standard exponential distribution). Therefore by considering $\mathcal{E}_{\tilde{U}}\{\exp(\theta \operatorname{tr} \tilde{A}\tilde{U}\tilde{B}\tilde{U}^*)\}$ where $\tilde{A} = \operatorname{diag}(\alpha_1, \ldots, \alpha_k)$, $\tilde{B} = \operatorname{diag}(\beta_1, \ldots, \beta_k)$, and $\tilde{U}$ is composed of independent standard complex normal variables, we obtain

Theorem 9.   *(Theorem 3.4.1)*

$$(20) \quad \prod_{i,j}^{k}(1 - \theta\alpha_i\beta_j)^{-1} = \sum_{n=0}^{\infty}(\theta^n/n!)\sum_{p\in P_n}\tilde{d}_p\,\tilde{Z}_p(\tilde{A})\,\tilde{Z}_p(\tilde{B}),$$

where $\tilde{d}_p$ is determined by

$$(21) \quad (\operatorname{tr} \tilde{A})^n = \sum_{p\in P_n}\tilde{d}_p\,\tilde{Z}_p(\tilde{A}).$$

Coefficients of $\tilde{Z}_p$ can be obtained as in the real case, namely (i) compare the coefficients of θ^n in both sides of (20), (ii) express the left hand side as a quadratic form in M_p or U_p, (iii) do the triangular decomposition to the resulting positive definite symmetric matrix of coefficients. Now it will be shown in the next section that the Schur functions S_p satisfy the same generating function (20) and S_p is a linear combination of lower order M_q's ($S_p = \sum_{q\leq p} a_{pq}M_q$). Therefore the Schur functions agree with the complex zonal polynomials by the uniqueness of the triangular decomposition of a positive definite symmetric matrix.

§ 5.3 SCHUR FUNCTIONS

In this section we give a definition of the Schur functions and show that they coincide with complex zonal polynomials by using Saw's generating function. In terms of the Schur functions Saw's generating function is given in Section 1.4 of Macdonald (1979) or in Section 7.6 of Weyl (1946).

Let $p = (p_1, \ldots, p_\ell) \in P_n$. The Schur function $S_p(x_1, \ldots, x_k)$ $(k \geq \ell)$

is defined by

(1)
$$S_p(x_1,\ldots,x_k) = \det(x_j^{p_i+k-i})_{1\le i,j\le k} / \det(x_j^{k-i})_{1\le i,j\le k}$$

$$= \begin{vmatrix} x_1^{p_1+k-1} & \cdots & x_k^{p_1+k-1} \\ x_1^{p_2+k-2} & \cdots & x_k^{p_2+k-2} \\ \vdots & & \vdots \\ x_1^{p_k} & \cdots & x_k^{p_k} \end{vmatrix} \div \begin{vmatrix} x_1^{k-1} & \cdots & x_k^{k-1} \\ \vdots & & \vdots \\ x_1 & \cdots & x_k \\ 1 & \cdots & 1 \end{vmatrix}.$$

If $k < \ell$ we define $S_p(x_1,\ldots,x_k) = 0$. See formula (3.1), Section 1.3 of Macdonald (1979). It is given in formula (35) of James (1964). In Weyl (1946) it is introduced as the primitive character of the unitary group and as the polynomial character of the general linear group (see Sections 7.5 and 7.6 of Weyl (1946)).

Note that the denominator of (1) is the *Vandermonde determinant*

(2)
$$\det(x_j^{k-i}) = \prod_{i<j}(x_i - x_j).$$

Clearly the numerator has $(x_i - x_j)$ as a factor because if $x_i = x_j$ then $\det(x_j^{p_i+k-i}) = 0$. Running (i,j) over all pairs we see that the numerator has the Vandermonde determinant as a factor. Furthermore if x_i and x_j are interchanged then both the numerator and the denominator change the sign and the ratio remains the same. Therefore $S_p(x_1,\ldots,x_k)$ is a symmetric polynomial in x_i's. It is easy to see that it is homogeneous of degree $|p|$. Now we want to show that

(3)
$$S_p(x_1,\ldots,x_k,0) = S_p(x_1,\ldots,x_k).$$

The last column of $(x_j^{p_i+k+1-i})_{1\le i,j\le k+1}$ is

$$(x_{k+1}^{p_1+k},\ldots,x_{k+1}^{p_k+1},x_{k+1}^{p_{k+1}+1})'.$$

If $x_{k+1} = 0$ it reduces to $(0,\ldots,0,1)'$. (Note that $p_{k+1} = 0$ by definition.) Hence if $x_{k+1} = 0$ then $\det(x_j^{p_i+k+1-i}) = (\prod_{j=1}^k x_j)\det(x_j^{p_i+k-i})$, the right

§ 5.3 Schur functions

hand side being the determinant of the $k \times k$ principal minor of the matrix on the left hand side. Similarly $\det(x_j^{k+1-i}) = (\prod_{j=1}^{k} x_j) \det(x_j^{k-i})$. Therefore we have (3) and in general by induction

(4) $$S_p(x_1, \ldots, x_k, 0 \ldots, 0) = S_p(x_1, \ldots, x_k).$$

This shows that $S_p \in V_n$. Now let us look at the highest monomial in S_p of the form $a x_1^{q_1} \cdots x_k^{q_k}$ $((q_1, \ldots, q_k) \in P_n)$. In $\det(x_j^{p_i+k-i})$ and $\det(x_j^{k-i})$ the similar terms are obtained by the products of the diagonal elements. They are

$$x_1^{p_1+k-1} x_2^{p_2+k-2} \cdots x_k^{p_k}, \qquad x_1^{k-1} \cdots x_{k-1}$$

respectively. From $S_p(x_1, \ldots, x_k) \det(x_j^{k-i}) = \det(x_j^{p_i+k-i})$ we obtain

$$(a x_1^{q_1} \cdots x_k^{q_k})(x_1^{k-1} \cdots x_{k-1}) = x_1^{p_1+k-1} \cdots x_k^{p_k}$$

Therefore $a = 1$ and $q = (q_1, \ldots, q_k) = (p_1, \ldots, p_k) = p$. We summarize these results in a lemma.

Lemma 1.

(5) $$S_p = M_p + \sum_{q < p} a_{pq} M_q,$$

and $\{ S_p, p \in P_n \}$ *forms a basis of* V_n.

Now we prove the following.

Lemma 2.

(6) $$\prod_{i,j}^{k}(1 - \theta x_i y_j)^{-1} = \sum_{n=0}^{\infty} \theta^n \sum_{p \in P_n} S_p(x_1, \ldots, x_k) S_p(y_1, \ldots, y_k).$$

Proof. Replacing x_i by θx_i we can assume $\theta = 1$ without loss of generality. We prove (6) in the following form:

(7) $$\frac{\det(x_j^{k-i}) \det(y_j^{k-i})}{\prod_{i,j=1}^{k}(1 - x_i y_j)} = \sum_{n=0}^{\infty} \sum_{p \in P_n} \det(x_j^{p_i+k-i}) \det(y_j^{p_i+k-i}).$$

We recognize the left hand side of (7) as Cauchy's determinant, i.e.

$$(8) \quad \frac{\det(x_j^{k-i})\det(y_j^{k-i})}{\prod(1-x_iy_j)} = \det\left(\frac{1}{1-x_iy_j}\right)_{1\leq i,j\leq k}$$

See Lemma 2 of Anderson and Mentz (1977) for example. To prove this directly consider the matrix on the right hand side. Now subtract appropriate multiples of the first row from other rows so that each element in the first column is converted to 0, except for that in the first row. Then the rest of the right hand matrix becomes

$$(9) \quad \frac{1}{1-x_iy_j} - \frac{1}{1-x_1y_j}\frac{1-x_1y_1}{1-x_iy_1} = \frac{x_1-x_i}{1-x_iy_1}\frac{y_1-y_j}{1-x_1y_j}\frac{1}{1-x_iy_j}$$

The first two factors of the right hand side come out of the determinant as common factors and (8) is now proved by induction on dimensionality. Now to derive the right hand side of (7) from the Cauchy's determinant, expand $(1-x_iy_j)^{-1}$ as $1+x_iy_j+x_i^2y_j^2+\ldots$ for every element of the matrix and then expand the determinant. Consider the term of the form $cx_1^{\ell_1}\cdots x_k^{\ell_k}$, $\ell_1 \geq \cdots \geq \ell_k$. This term arises as follows. Take ℓ_i-th power term in each element of the i-th row, $i=1,\ldots,k$. Then $x_i^{\ell_i}$ comes out as a common factor and we obtain $x_1^{\ell_1}\cdots x_k^{\ell_k}\det(y_j^{l_i})$. Collecting permuted terms in x's we have $\det(x_j^{\ell_i})\det(y_j^{\ell_i})$. Therefore Cauchy's determinant can be expanded as

$$(10) \quad \sum_{\ell_1\geq\cdots\geq\ell_k} \det(x_j^{\ell_i})\det(y_j^{\ell_i}).$$

Now if $\ell_i=\ell_{i+1}$ for some i, then $\det(x_j^{\ell_i})=0$. Hence this summation is actually over the set $\{(\ell_1,\ldots,\ell_k):\ell_1>\cdots>\ell_k\}$. Now letting $\ell_i=p_i+k-i$, $i=1,\ldots,k$ the summation becomes over all partitions $p=(p_1,\ldots,p_k)$. This proves the lemma. ∎

This proof has been adapted from p.202, Section 7.6 of Weyl (1946).

Comparing (5.2.20) and (6) we have

$$(11) \quad \sum_{p\in P_n}(\tilde{d}_p/n!)\tilde{Z}_p(A)\tilde{Z}_p(B) = \sum_{z\in P_n} S_p(A)S_p(B)$$

Now when expressed with respect to the basis $\{M_q\}$, both $\tilde{Z}_p$ and S_p are linear combinations of M_q with $q \leq p$. Therefore if (11) is expressed in terms of M_q's, then two sides of (11) give the same triangular (: lower times upper) decomposition of a positive definite coefficient matrix. By the uniqueness of the triangular decomposition of a symmetric positive definite matrix we have $S_p = c_p \,_1\tilde{\mathcal{Y}}_p$ for some c_p. Comparing the leading coefficient (see (5)) we obtain $c_p = 1$. Furthermore considering the leading coefficient of $\tilde{Z}_p$ we obtain $\tilde{d}_p \,_1\tilde{b}_p^2 = n!$, $n = |p|$. This was mentioned at the end of Section 4.2.

We have proved

Theorem 1.

(12) $$S_p = \,_1\tilde{\mathcal{Y}}_p \quad \text{and} \quad \tilde{d}_p \,_1\tilde{b}_p^2 = n! \quad \text{where} \quad n = |p|.$$

There are three more determinantal expressions stated in James (1964). One involving elementary symmetric functions (formula(37) in James (1964)) is found in (2.9') of Macdonald. One involving "complete symmetric functions" (formula(36) of James (1964)) is given in (3.4) of Macdonald. Formula (38) in James (1964) is not given in Macdonald.

Elementary symmetric functions and complete symmetric functions of the roots of a matrix are relatively easy to calculate. Determinants can be evaluated easily by computer as well. Hence from the viewpoint of numerical computation these determinantal expressions seem to be all we have to know. Namely we do not need to know the coefficients of M_p or U_p etc. to evaluate the Schur function. It might be worthwhile to look for an analogue of this for the real zonal polynomial. Another possibility is to express the real zonal polynomials in terms of the Schur functions.

§ 5.4 RELATION BETWEEN THE REAL AND THE COMPLEX ZONAL POLYNOMIALS

We finish this chapter by discussing some results which we were unable to derive by our elementary approach. James (1964) gives the following formula relating the complex and the real zonal polynomials:

$$\text{(1)} \qquad \frac{Z_p(XX')}{Z_p(I_k)} = \mathcal{E}_H S_{2p}(XH),$$

where the $k \times k$ H has the uniform distribution of orthogonal matrices, $p = (p_1,\ldots,p_\ell) \in P_n$, and $2p = (2p_1,\ldots,2p_\ell) \in P_{2n}$. (Formula (34) in James (1964).) Furthermore he states

$$\text{(2)} \qquad \mathcal{E}_H S_p(XH) = 0,$$

if one or more parts of p is odd. (Formula (40)). See also Theorem 12.11.6 and Remark 12.11.11 in Farrell (1976).

Given these results we can evaluate d_p in (3.4.1) as follows. First note that by replacing H by U where U is composed of independent standard (real) normal variables we obtain

$$\text{(3)} \qquad \begin{aligned} \mathcal{E}_U S_{2p}(XU) &= \mathcal{E}_{T,H} S_{2p}(XTH) \\ &= \mathcal{E}_T Z_p(XTT'X')/Z_p(I_k) \\ &= Z_p(XX'). \end{aligned}$$

By (5.2.21) and (5.3.12)

$$\text{(4)} \qquad \begin{aligned} (\operatorname{tr} A)^{2n} &= \sum_{p \in P_{2n}} \tilde{d}_p \tilde{Z}_p(A) \\ &= \sum_{p \in P_{2n}} \tilde{d}_p\, _1\tilde{b}_p\, _1\tilde{\mathcal{Y}}_p(A) \\ &= \sum_{p \in P_{2n}} (2n)!\, _1\tilde{b}_p^{-1}\, _1\tilde{\mathcal{Y}}_p(A). \end{aligned}$$

Now let $A = \operatorname{diag}(\alpha_1,\ldots,\alpha_k)$ and replace A by AU. In this case

$$\operatorname{tr} AU = \sum_{i=1}^k \alpha_i u_{ii} \sim \mathcal{N}(0, \sum \alpha_i^2).$$

Hence

$$\text{(5)} \qquad \begin{aligned} \mathcal{E}_U(\operatorname{tr} AU)^{2n} &= 1 \cdot 3 \cdots (2n-1)(\sum \alpha_i^2)^n \\ &= \frac{(2n)!}{2^n n!}(\operatorname{tr} AA')^n. \end{aligned}$$

On the other hand by (3) and (2)

$$(6) \qquad \mathcal{E}_U \sum_{p \in P_{2n}} (2n)! \, {}_1\tilde{b}_p^{-1} \, {}_1\tilde{y}_p(AU) = \sum_{p \in P_n} (2n)! \, {}_1\tilde{b}_{2p}^{-1} \, Z_p(AA').$$

Hence comparing (4), (5), and (6) we obtain

$$(\operatorname{tr} AA')^n = \sum_{p \in P_n} {}_1\tilde{b}_{2p}^{-1} \, 2^n n! \, Z_p(AA')$$

or

$$(7) \qquad d_p = 2^n n! \, {}_1\tilde{b}_{2p}^{-1}.$$

Now (5.2.19) gives (3.4.12).

References

Anderson, T.W. (1946). The noncentral Wishart distribution and certain problems of multivariate statistics. *Ann. Math. Statist.*, **17,** 409-431.

Anderson, T.W. (1958). *An introduction to multivariate statistical analysis.* Wiley, New York.

Anderson, T.W. and Girshick, M.A. (1944). Some extensions of the Wishart distribution. *Ann. Math. Statist.*, **15,** 345-357.

Anderson, T.W. and Mentz, R.P. (1977). The generalized variance of a stationary autoregressive process. *J. Multivariate Anal.*, **7,** 584-588.

Brillinger, D.R. (1975). *Time series: data analysis and theory.* Holt, Rinehart and Winston, New York.

Chikuse, Y. (1981). Distributions of some matrix variates and latent roots in multivariate Behrens-Fisher discriminant analysis. *Ann. Statist.*, **9,** 401-407.

Constantine, A.G. (1963). Some noncentral distribution problems in multivariate analysis. *Ann. Math. Statist.*, **34,** 1270-1285.

Constantine, A.G. (1966). The distribution of Hotelling's generalized T_0^2. *Ann. Math. Statist.*, **37,** 215-225.

Crowther, N.A.S. and DeWaal, D.J. (1973). On the distribution of a generalised positive semidefinite quadratic form of normal vectors. *South African Statist. J.*, **7,** 119-127.

David, F.N., Kendall, M.G., and Barton, D.E. (1966). *Symmetric function and allied tables*. Cambridge University Press.

Davis, A.W. (1979). Invariant polynomials with two matrix arguments extending the zonal polynomials: applications to multivariate distribution theory. *Ann. Inst. Statist. Math.*, **31,** 465-485.

Davis, A.W. (1980). Invariant polynomials with two matrix arguments, extending the zonal polynomials. In *Multivariate Analysis-V*, P.R. Krishnaiah, ed. North-Holland.

Davis, A.W. (1981). On the construction of a class of invariant polynomials in several matrices, extending the zonal polynomials. *Ann. Inst. Statist. Math.*, **33,** 297-313.

DeWaal, D.J. (1970). Distributions connected with a multivariate beta statistics. *Ann. Math. Statist.*, **41,** 1091-1095.

DeWaal, D.J. (1972). An asymptotic distribution of noncentral multivariate Dirichlet variates. *South African Statist. J.*, **6,** 31-40.

DeWaal, D.J. (1973). On the elementary symmetric functions of the Wishart and correlation matrices. *South African Statist. J.*, **7,** 47-60.

Farrell, R.H. (1976) *Techniques of multivariate calculation.* Vol.520, Lecture Notes in Mathematics, Springer-Verlag, Berlin.

Farrell, R.H. (1980). Calculation of complex zonal polynomials. In *Multivariate Analysis-V* (P.R.Krishnaiah, ed.), North-Holland, pp. 301-320.

Gantmacher, F.R. (1959). *The theory of matrices, Vol. I*. Chelsea, New York.

Goodman, N.R. (1963). Statistical analysis based on a certain multivariate complex Gaussian distribution (an introduction). *Ann. Math. Statist.*, **34,** 152-177.

Halmos, P.R. (1974). *Measure Theory.* Springer, New York.

Hayakawa, T. (1966). On the distribution of a quadratic form in a multivariate normal sample. *Ann. Inst. Statist. Math.*, **18,** 191-201.

Hayakawa, T. (1967). On the distribution of the maximum latent root of a positive definite symmetric random matrix. *Ann. Inst. Statist. Math.*, **19,** 1-17.

Hayakawa, T. (1969). On the distribution of the latent roots of a positive definite random symmetric matrix I. *Ann. Inst. Statist. Math.*, **21**, 1-21.

Hayakawa, T. and Kikuchi, Y. (1979). The moments of a function of traces of a matrix with a multivariate symmetric normal distribution. *South African Statist. J.*, **13**, 71-82.

Helgason, S. (1962). *Differential geometry and symmetric spaces*. Academic Press, New York.

Herz, C.S. (1955). Bessel functions of matrix argument. *Ann. Math.*, **61**, 474-523.

Hua, L.K. (1959). *Harmonic analysis of functions of several complex variables in the classical domains*. Moscow. (English translation: Amer. Math. Soc., Providence, R.I.)

James, A.T. (1954). Normal multivariate analysis and the orthogonal group. *Ann. Math. Statist.*, **25**, 40-75.

James, A.T. (1955a). The noncentral Wishart distribution. *Proc. Roy. Soc. (London), A*, **229**, 364-366.

James, A.T. (1955b). A generating function for averages over the orthogonal group. *Proc. Roy. Soc. (London), A*, **229**, 367-375.

James, A.T. (1960). The distribution of the latent roots of the covariance matrix. *Ann. Math. Statist.*, **31**, 151-158.

James, A.T. (1961a). Zonal polynomials of the real positive definite symmetric matrices. *Ann. Math.*, **74**, 456-469.

James, A.T. (1961b). The distribution of noncentral means with known covariances. *Ann. Math. Statist.*, **32**, 874-882.

James, A.T. (1964). Distribution of matrix variates and latent roots derived from normal samples. *Ann. Math. Statist.*, **35**, 475-501.

James, A.T. (1968). Calculation of zonal polynomial coefficients by use of the Laplace-Beltrami operator. *Ann. Math. Statist.*, **39**, 1711-1718.

James, A.T. (1973). The variance information manifold and the functions on it. In *Multivariate Analysis-III* (ed. P.R.Krishnaiah). Academic Press, New York.

Johnson, N.L. and Kotz, S. (1972). *Distributions in statistics: continuous multivariate distributions.* Wiley, New York.

Kates, L.K. (1980). Zonal polynomials. Ph.D. dissertation, Princeton University.

Khatri, C.G. (1966). On certain distribution problems based on positive definite quadratic functions in normal vectors. *Ann. Math. Statist.,* **37**, 468-479.

Khatri, C.G. (1967). Some distribution problems connected with the characteristic roots of $S_1 S_2^{-1}$. *Ann. Math. Statist.,* **38**, 944-948.

Khatri, C.G. (1972). On the exact finite series distribution of the smallest or the largest root of matrices in three situations. *J. Multivariate Anal.,* **2**, 201-207.

Khatri, C.G. and Pillai, K.C.S. (1968). On the noncentral distribution of two test criteria in multivariate analysis of variance. *Ann. Math. Statist.,* **39**, 215-226.

Krishnaiah, P.R. and Chang, T.C. (1971). On the exact distribution of the smallest root of the Wishart matrix using zonal polynomials. *Ann. Inst. Statist. Math.,* **23**, 293-295.

Kshirsagar, A.M. (1959). Bartlett decomposition and Wishart distribution. *Ann. Math. Statist.,* **30**, 239-241.

Kushner, H.B. and Meisner, M. (1980). Eigenfunctions of expected value operators in the Wishart distribution. *Ann. Statist.,* **8**, 977-988.

Kushner, H.B., Lebow, A., and Meisner, M. (1981). Eigenfunctions of expected value operators in the Wishart distribution, II. *J. Multivariate analysis,* **11**, 418-433.

Lehmann, E.L. (1959). *Testing statistical hypotheses.* Wiley, New York.

Littlewood, D.E. (1950). *The theory of group characters and matrix representations of groups.* Clarendon, Oxford.

Loomis, L.H. (1953). *An introduction to abstract harmonic analysis.* D. Van Nostrand, New York.

Macdonald, I.G. (1979). *Symmetric functions and Hall polynomials.* Oxford Mathematical Monographs, Clarendon Press, Oxford.

Mardia, K.V., Kent, J.T., and Bibby, J.M. (1979). *Multivariate Analysis.* Academic Press, New York.

Marshall, A.W. and Olkin, I. (1979). *Inequalities: theory of majorization and its applications.* Academic Press, New York.

Mirsky, L. (1955). *An introduction to linear algebra.* Oxford University Press.

Muirhead, R.J. (1982). *Aspects of multivariate statistical theory.* Wiley, New York.

Nachbin, L. (1965). *The Haar integral.* Litton Education Pub., Inc.

Nagarsenker, B.N. (1978). Nonnull distribution of some statistics associated with testing for the equality of two covariance matrices. *J. Multivariate Anal.*, **8**, 396-404.

Olkin, I. and Perlman, M.D. (1980). Unbiasedness of invariant tests for MANOVA and other multivariate problems. *Ann. Statist.*, **8**, 1326-1341.

Parkhurst, A.M. and James, A.T. (1974). Zonal polynomials of order 1 through 12. in *Selected tables in mathematical statistics*, Vol.2. I.M.S. publications.

Pillai, K.C.S. (1967). On the distribution of the largest root of a matrix in multivariate analysis. *Ann. Math. Statist.*, **38**, 616-617.

Pillai, K.C.S. (1975). The distribution of the characteristic roots of $S_1 S_2^{-1}$ under violations. *Ann. Statist.*, **3**, 773-779.

Pillai, K.C.S. (1976). Distributions of characteristic roots in multivariate analysis. Part I. Null distributions. *Can. J. Statist.*, **4**, 157-184.

Pillai, K.C.S. (1977). Distributions of characteristic roots in multivariate analysis. Part II. Non-null distributions. *Can. J. Statist.*, **5**, 1-62.

Pillai, K.C.S., Al-Ani, S., and Jouris, G.M. (1969). On the distributions of the ratios of the roots of covariance matrix and Wilks' criterion for tests of three hypotheses, *Ann. Math. Statist.*, **40**, 2033-2040.

Pillai, K.C.S. and Nagarsenker, B.N. (1971). On the distribution of the sphericity test criterion in classical and complex normal populations having unknown covariance matrix. *Ann. Math. Statist.* **42**, 764-767.

Pillai, K.C.S. and Nagarsenker, B.N. (1972). On the distribution of a class of statistics in multivariate analysis. *J. Multivariate Anal.*, **2**, 96-114.

Pillai, K.C.S. and Sugiyama, T. (1969). Noncentral distributions of the largest latent roots of three matrices in multivariate analysis. *Ann. Inst. Statist. Math.*, **21**, 321-327.

Saw, J.G. (1970). The multivariate linear hypothesis with nonnormal errors and a classical setting for the structure of inference in a special case. *Biometrika*, **57**, 531-535.

Saw, J.G. (1977). Zonal polynomials: an alternative approach. *J. Multivariate Analysis*, **7**, 461-467.

Shah, B.H. (1970). Distribution theory of a positive definite quadratic form with matrix argument. *Ann. Math. Statist.*, **41**, 692-697.

Shorrock, R.W. and Zidek, J.V. (1976). An improved estimator of the generalized variance. *Ann. Statist.*, **4**, 629-638.

Srivastava, M.S. (1968). On the distribution of a multiple correlation matrix: noncentral multivariate beta distributions. *Ann. Math. Statist.*, **39**, 227-232.

Srivastava, M.S. and Khatri, C.G. (1979). *An introduction to multivariate statistics*. North Holland, New York.

Stewart, G.W. (1973). *Introduction to matrix computations*. Academic Press, New York.

Subrahmaniam, K. (1976). Recent trends in multivariate normal distribution theory: on the zonal polynomials and other functions of matrix argument. *Sankhya A*, **38**, Part 3, 221-258.

Sugiura, N. (1971). Note on some formulas for weighted sums of zonal polynomials. *Ann. Math. Statist.*, **42**, 768-772.

Sugiura, N. (1973). Derivatives of the characteristic root of a symmetric or a Hermitian matrix with two applications in multivariate analysis. *Communications in Statistics*, **1**, 393-417.

Sugiura, N. and Fujikoshi, Y. (1969). Asymptotic expansions of the non-null distributions of the likelihood ratio criteria for multivariate linear hypothesis and independence. *Ann. Math. Statist.*, **40**, 942-952.

Sugiyama, T. (1966). On the distribution of the largest latent root and the corresponding latent vector for principal component analysis. *Ann. Math. Statist.*, **37**, 995-1001.

Sugiyama, T. (1967a). On the distribution of the largest latent root of the covariance matrix. *Ann. Math. Statist.*, **38**, 1148-1151.

Sugiyama, T. (1967b). Distribution of the largest latent root and the smallest latent root of the generalized B statistic and F statistic in multivariate analysis. *Ann. Math. Statist.*, **38**, 1152-1159.

Towber, J. (1981). Personal communication.

Weyl, H. (1946). *The classical groups*. Princeton University Press, Princeton, N.J.

Wijsman, R.A. (1959). Application of a certain representation of the Wishart matrix. *Ann. Math. Statist.*, **30**, 597-601.

Zidek, J. (1978). Deriving unbiased risk estimators of multinormal mean and regression coefficient estimators using zonal polynomials. *Ann. Statist.*, **6**, 769-782.